Beekeeping
A Beginner's Guide

Other books by Spring Hill, an imprint of How To Books:

HOW TO GROW YOUR OWN FRUIT AND VEG
A week-by-week guide to wild-life friendly fruit and vegetable gardening
JOE HASHMAN

MAKING CIDER, MEAD, PERRY AND FRUIT WINES
Recipes and how to use them
CRAIG HUGHES

PLANTS AND PLANTING PLANS FOR A BEE GARDEN
How to design beautiful borders that will attract bees
MAUREEN LITTLE

THE BEE GARDEN
How to create or adapt a garden that will attract and nurture bees
MAUREEN LITTLE

THE BEE KEEPER'S FIELD GUIDE
A pocket guide to the health and care of bees
DAVID CRAMP

Write or phone for a catalogue to:

How To Books
Spring Hill House
Spring Hill Road
Begbroke
Oxford
OX5 1RX
Tel. 01865 375794

Or email: info@howtobooks.co.uk

Visit our website www.howtobooks.co.uk to find out more about us and our books.

Like our Facebook page **How To Books & Spring Hill**

Follow us on **Twitter** @**Howtobooksltd**

Read our books online www.howto.co.uk

How to keep bees and develop your interest further

Beekeeping
A Beginner's Guide

David Cramp

SPRING HILL

Published by Spring Hill,
an imprint of How To Books Ltd,
Spring Hill House, Spring Hill Road,
Begbroke, Oxford OX5 1RX
Tel: (01865) 375794. Fax: (01865) 379162
info@howtobooks.co.uk
www.howtobooks.co.uk

British Library Cataloguing in Publication Data
A catalogue record for this book is available from the British Library

ISBN: 978 1 905862 87 0

Cover design by Mousemat Design Ltd
Produced for How To Books by Deer Park Productions, Tavistock
Typeset by TW Typesetting, Plymouth, Devon
Printed and bound by Graficas Cems, Villatuerta (Spain)

Contents

Preface ix

Introduction xi

1. Why Become a Beekeeper? 1

What can beekeeping do for you? 1

For the keen gardener 2

The honey harvest 3

Becoming a livestock farmer 4

Becoming a bee vet 5

Helping the planet 5

Learning about the history of the world 7

Helping to characterise our age 7

Where do beekeepers really come in? 7

But what else is there to beekeeping? 8

2. Frequently Asked Questions about Beekeeping 15

Bee stings? 15

Is it worth signing up for a beekeeping course? 17

What equipment will I need and how much will it cost? 18

What are the health requirements? 23

How much time and effort is required? 23

Where could I keep my bees? 23

How will children, neighbours and pets be affected? 28

Do I need permission to keep bees, even in my garden? 28

Is insurance required? 29

What support will I get and from whom? 29

Reading 30

3. Hives 31

So how many hives should I start with? 31

What type of hives should I use? 32

Do the bees care what type of hives they occupy? 41

Requirement for proper hive research   42

Warre or other top bar hive?   43

4. Bees   44

Seek local advice   45

The hive occupants   46

Where to get bees from   47

What are you actually getting? The bees in the hive 49

Some taxonomy and basic bee biology   53

Local bees or any old bee?   55

Bee talk in the dark   55

The bee – a very specialised creature   57

Bee vision – a different reality   57

The bee in form and function   58

Pheromones in control   58

Conclusion   59

5. Spring – Your Bees Arrive   60

The bee year   60

What you have already done   60

Your bees arrive   61

Scenario 1: Existing hives   61

Scenario 2: A swarm   62

Scenario 3: The nucleus   65

You have your bees!   67

First steps for spring   67

Inspecting your bees   68

Are there any signs of disease?   76

Are there sufficient stores of honey?   76

What about pollen?   77

The inspection   78

Keep a written record   79

Interpreting honey flows   80

Conclusion and summary   80

INTERNAL INSPECTION WORK SHEET   81

6. Spring to Summer   82
   Varroa destructor   82
   Swarming   83
   Methods and manipulations to help prevent swarming   86
   The artificial swarm   88
   Conclusion   93

7. Summer – Hive Problems and the Upcoming Harvest   94
   Regular inspections   94
   The external inspection   95
   Robbing   98
   Dealing with aggressive colonies   100
   A new queen?   100
   The laying worker   101
   Spray poisoning and moving hives   102
   Moving bees for short and long journeys   103
   Conclusion   104

8. Summer into Autumn – Harvest Day and Post-Harvest Work   105
   Planning is everything   105
   Your extraction plant!   105
   What equipment do I need?   106
   Preparing the hives for harvesting   109
   Hard working slaves   109
   How to set up your kitchen   110
   Clearing the bees   110
   The extraction process   111
   Storing honey   115
   Selling honey   115
   Testing honey   117
   What about the wet frames?   117
   Wax comb storage   118
   Varroa again   119
   Conclusion   119

9. Winter and Beyond   120
    The year past   120
    Moving towards winter and pre-winter tasks   121
    Other wintering tasks   123
    Arrival of winter   123
    Winter projects   126
    And so through to spring and the need for pollen   127
    Arrival of spring   128
    Conclusion   130

10. Bee Diseases and Pests   132
    How to tell if a colony has a disease   132
    What differences could there be and are they significant?   133
    Pests and diseases   134
    Brood diseases   135
    Adult bee diseases   137
    Bee pests   139
    Conclusion   144

Appendices: Useful Information for Beekeepers   146
    Appendix 1: A Beekeeping Calendar   146
    Appendix 2: What to do With it All   150
    Appendix 3: Taking it Further   157
    Appendix 4: Glossary of Terms   164

Index   173

Preface

I wrote this book to encourage readers to take up beekeeping or at least to learn about these fascinating and essential creatures. When talking to people about bees, or when giving talks to a variety of organisations, I was always surprised not only by their genuine interest in the subject, but also at their lack of knowledge. This was amply shown by their reactions . . . 'I simply didn't realise . . .', 'that's just amazing . . .', 'I really didn't know how important they are . . .' and so on which were all typical remarks. Because of this and with the publisher's encouragement, I decided to write a beginner's guide to beekeeping which not only teaches readers how to keep bees in an easily understood manner, but also explains just how important bees are to us all.

One of the features of this book is the length it goes to, to answer questions that interested people may ask long before they even think about taking up beekeeping, and for much of the information on this I must thank Ron Levin, a hobby beekeeper belonging to the Harrow Beekeepers' Association who gave me much advice from a beginner's perspective. He told me what people really wanted to know about before they decide to start beekeeping and the first part of the book is devoted to that subject.

I hope that this book will introduce very many people to bees even if they don't start beekeeping and I hope that it will remind people that bees are an essential component of our ecological and economic landscape and critical to our own welfare.

Photo/picture credits:
I must also gratefully acknowledge the following photographs;

Photo 22: Peter Mcfadden, Conwy BKA, UK

Photos 28, 29, 35, 49, 51 and 52: Food and Environment Research Agency (FERA) reproduced under the terms of the Click-Use licence No: C2010002634

Photos 20 and 34: Colin Eastham

Figures 6a and 6b: Reproduced under the terms of Creative Commons Attribution SA 3 licence

Figure 7: Reproduced under the terms of Creative Commons Attribution 2.5 licence

Introduction

This book offers two things: firstly, an introduction to bees and beekeeping with an explanation of why beekeeping is such a brilliant pastime to take up, and secondly, a simple, basic beekeeping primer for the complete beginner. Beekeeping isn't just a hobby; it is also a window into the natural world. Truly natural! Bees are totally wild creatures. They are pollinators and there are arguably fewer creatures more important to the health of our planet. They will go through their entire lives oblivious to you as a beekeeper except perhaps to regard you as just another possible predator that needs to be deterred.

Beekeeping is also a multi-faceted pastime – and for some a career – that whether on a large or small scale – offers so many opportunities and learning possibilities that it may be time to introduce these other aspects of beekeeping to readers who may never have thought about beekeeping before. Beekeepers themselves are vital simply because they keep pollinating insects – and that is of increasing importance in today's world. Why is that? Let's look at what the executive director of UNEP said in 2011:

> The way humanity manages or mismanages its nature based assets, including pollinators will in part define our collective future in the 21st century. Human beings have fabricated the illusion that in the 21st century they have the technological prowess to be independent of nature. Bees underline the reality that we are more, not less dependant on nature's services in a world of close to seven billion people.

Do you want to take part in managing these assets – whilst at the same time enjoying a thoroughly interesting and wide ranging hobby? I hope so. Beekeeping is different and beekeepers really stand out from the crowd.

One of the aims of this book then is to introduce this 'different look' on beekeeping to those who thought it was just about honey or for those who may not have thought about it at all. For some, the very thought of keeping a bunch of wild, stinging insects as livestock sounds insane – honey or not – but in fact when you better understand what bees and beekeeping are all about and when you discover what beekeeping can do for you and all of the endless avenues that you can explore, you may look on it all in a different light.

For those who accept the challenge of taking up this ecologically sound hobby, the basic premise of the book is that the local beekeeping association forms the bedrock of beekeeping knowledge and skills for all beekeepers. With this in mind I urge anyone thinking of starting beekeeping to contact their local association without hesitation. Without this essential support group at your side, your beekeeping will be less enjoyable, less certain and less knowledge based. Use this book and the reservoir of skills and knowledge of association beekeepers to advance your beekeeping interest. A book can only tell you so much. Without support, your hobby will be the poorer for it.

The essential supporting role of the local beekeeping association is explained and the support of government agencies and agricultural concerns are also outlined so as to give a reassuring picture to the new or prospective beekeeper.

1
Why Become a Beekeeper?

The honey bee uses the most complex symbolic language of any
animal on earth, outside of the primate family – and it is the only
animal language that is understood by man. Want to learn it?

The question 'Why become a beekeeper?' is not really an unusual question these days because most people have never really thought much about beekeeping, and if they did it is often because of some elderly relative – an uncle or aunt or grandparent who used to keep bees in the garden somewhere back in the dark ages. But in the modern world when you can get a range of honeys at a reasonable price from every small provision shop or supermarket, why bother to keep bees yourself. Many believe the myths that bees attack on sight; they sting everyone; they bother the neighbours if you live in town. They need looking after if you want honey and they look and act like wasps. And in believing all that, they legitimately ask 'so what on earth is the lure?' What else is there about beekeeping that can possibly attract anyone?

Let's take a look at the truly exciting menu of possibilities that opens up for anyone who keeps bees and discover just what this subject can offer you. This menu varies from simple beekeeping for honey through pollination and food production studies, microscopy, clinical studies, stamp collecting, book collecting, poetry, cooking, brewing, making polishes and cosmetics, photography, evolution studies and planetary history, right through to helping the ecology of the planet. It really is quite something – and something that you will never tire of.

What can beekeeping do for you?

People take up beekeeping for many reasons. Some because they want to engage with a truly wild creature and open a window into the natural world for themselves and their families; others because they genuinely want to help the local ecology. Two beekeepers I know, both busy executives simply attend

The author with his first honey harvest.

to their bees when necessary and for the rest of the time just sit near the hives watching the bees at work and play whilst relaxing over a glass or two of wine. The bees keep the honey and the beekeepers de-stress with nature! Even Michelle Obama put hives on the White House lawn, following the trend in urban beekeeping which is becoming a really big movement now. This is healthy because being a beekeeper forces you to slow down. Bees sense stress, and they sense trouble. You need to be relaxed and stress-free around them.

For the keen gardener

Keeping bees is a natural extension of gardening and completes the circle as far as a gardener is concerned. Growing plants and vegetables is a wonderful and very practical pastime, but without bees gardeners are only getting half the picture in learning about just how their flowers and vegetables grow. Beekeeping can be a very satisfying extension to their hobby and one which I am sure they would not want to do without once they start.

Keeping bees is a natural extension activity for many gardeners.

The honey harvest

For many, the honey harvest is a tangible expression of their beekeeping and it can be an exciting time. This is what they aim for and are interested in. Then, when this almost magical liquid is safely jarred or stored in tubs, the kitchen floor cleaned up and all the other associated mess sorted out, you can decide what to do with it all. Sell it at the local farmer's market! (Watch out for all the bureaucracy in this.) Exchange it for hay for the kids' horses; make mead out of it; use it in cookery; make lotions and creams from it, display it at honey shows – there are a million possibilities. One beekeeper gives it to his work colleagues for donations of money and then sends the money to his favourite charity. Most of his colleagues had never tasted anything but supermarket honey before and couldn't believe the taste sensation of pure, raw honey straight from the comb and filtered through the kitchen sieve – just get the big bits out such as the odd head or leg of bee!

Harvesting and using other products

The harvest doesn't only include honey however. Wax, pollen, royal jelly and propolis and even bee venom make up a sizeable part of the harvest for some

A typical small harvest of liquid honey and cut comb.

beekeepers who use these substances as a profitable addition to the honey harvest. Just imagine the fun and satisfaction of making your own brand face creams or scented candles or superior furniture polish. There are markets for all of these products and many beekeepers use wax as another offshoot craft altogether by making candles, soaps, wax products, polishes and creams, which all have a wonderful fragrance and make fantastic presents.

But it isn't only all about honey or other hive products

One comment that many new beekeepers make is that they didn't all take up beekeeping just to make honey. In fact for some, producing honey was a very low priority initially and only became a wonderful bonus later on when they found that not only could they produce their own honey but that it tasted like honey as well. So if it wasn't producing honey, what else tempted them into keeping thousands of stinging insects often in their back garden right next to the house?

Becoming a livestock farmer

Beekeeping can bring out the farmer in you! Bees are not pets to be kept. Despite appearances, bees have never become domesticated. Even with one hive, beekeeping is small scale livestock farming. It isn't simply a hobby or craft because you have a responsibility for a colony, or several colonies of bees that in return will be producing honey for you and pollinating the surrounding flora. Like any pro-active farmer you will instinctively start looking at weather forecasts and checking out flowering times of the local flowers. Studying the local flora which is (and the bees) food source will become automatic and studying the honey and pollen producing properties of the flowers will take up much of your thinking time. Is there an early source of pollen so essential for the spring build up? Will the early rain wash out all the nectar from the apple blossom? Will the unexpected heat dry out the wild flowers? These are all questions that every beekeeper thinks about on an almost daily basis and within a year you will begin to acquire an expert's knowledge of the properties of every flower in your surrounding area.

Becoming a bee vet

Again like any other farmer you will look out for livestock diseases and without realising it you will gradually become skilled at recognising disease symptoms and deciding how to treat them. Remember, there are no bee vets! You will instinctively learn about what harms your bees and when. You'll become an expert on local farming practices that use pesticides and herbicides and you'll know what to do about each threat. You'll have the active help of both the government and the local beekeeping organisation if you ask for help but it will be you who will call the initial shots. Treating for existing problems will become a routine and all the other responsibilities for livestock care will be yours. In short, there's a whole wealth of knowledge that you will gradually acquire, things like: meteorology, bee biology, bee pests and diseases, the government legislation, pollination, crops, botany, nutritional aspects, more government legislation and so on and so on. Don't be daunted by the list because it all comes naturally and gradually.

Helping the planet

In recent years there have been many reports, not only from the UK but also from many other countries, that our bees and other pollinators are rapidly disappearing and that governments are becoming increasingly concerned about the reasons for this. So why would governments need to get involved

This sunflower cannot produce seeds without pollination.

in what in most people's mind is 'just' a quaint hobby for a few mild eccentrics? Well, the answer to that question is that beekeeping isn't 'just' that at all! The government isn't interested in honey production (unless you sell it and then they want their tax). Bees pollinate crops and pasture and without them and other pollinators our agriculture suffers. Agricultural profits and employment suffers, our food supply suffers and governments throughout the world are getting very worried about it all. There are many reasons why this state of affairs has come about and we'll take a look at them in a future chapter but for the moment beekeeping is arguably one of the only hobbies or pastimes where an individual can actually make a genuine difference to this worrying situation. By keeping bees you are directly and immediately contributing to the health and welfare of the national agricultural system; the ecological environment and the health and safety of the planet. That sounds pretty impressive but do the facts support the statement? Can you actually make a difference and is it only agriculture that suffers if there are fewer beekeepers.

Learning about the history of the world

The honey bee is the product of well over 100 million years of evolution as a pollinator. It isn't the only pollinator around and indeed it is unable to pollinate some flowers, but it is hugely important especially for much of the food we eat. Pollination colours our world; feeds our world; enables our world to breathe and provides us all with the basic framework upon which all of our elementary life support systems depend.

Helping to characterise our age

Strong evidence suggests that we are wittingly or unwittingly causing the disappearance of the honey bee – and other pollinators – by the use of systemic insecticides of such potency that even undetectable doses can reduce the bees' immune system to such an extent that other pests and diseases can then cause their loss – a sort of 'bee aids'. Others will argue with that assessment but for whatever reason, the facts of the matter are that bees are disappearing in vast numbers and it is costing billions in lost food production as well as huge sums in researching the causes. Other pollinators are also disappearing as well and our 'life support system' is slowly dying off.

One commentator Michael McCarthy of the Independent newspaper said in 2011, 'This isn't just about bees – it affects everything'. And he is right. He also goes on to ask, 'How will we characterise our age? By the birth of the internet? The rise of China? The first black US president? Perhaps in all those ways. But we could also say, less obviously but perhaps more fundamentally, that ours is the age when the insects disappeared.' Can we ignore that?

Where do beekeepers really come in?

There are whole books written about saving the planet but can you contribute? By taking up beekeeping you are not only engaging with a truly wild creature, but as I mentioned earlier, you are immediately helping this dire situation to the tune of some 50,000 pollinators – and that's if you have only one hive! Remember the average worker bee will visit (and pollinate) 50 to 100 flowers during each foraging trip and they will make several trips per day. Add to this figure the efforts of their 20 or 30,000 sisters you can start to imagine the huge pollination task that they successfully achieve and that's only in one day from one colony! So you can make an enormous difference. In short, beekeepers have bees which keep agricultural and horticultural. systems going as well as our parks and gardens and the natural environment. And that's why you can very effectively help with our planetary health and is one of the main reasons why keeping bees is such a rewarding experience. You may not be out on the southern oceans saving the whales with Greenpeace or leading the charge against the destruction of the rain forests, but you will be acting just as effectively. You will become a small scale livestock farmer of inestimable value to us all. In the UK alone, The British Beekeepers' Association estimates that over 90 per cent of the UK's honeybee population belongs to the efforts of 17,000 amateur beekeepers. This is taken from the UK government's Food and Environment Research Agency (FERA) website:

> Honey bees make an important contribution to sustainable agriculture and the environment. The Government recognises the importance of a strong bee health programme in England to protect these benefits and takes very seriously any bio-security threat to the sustainability of the apiculture sector.

You can see then that beekeepers are important. And remember, you are also engaging in something that in this day and age has almost been forgotten about – stewardship of resources. Resources that will disappear if we are not careful.

Are you ready to join their ranks?

But what else is there to beekeeping?

The whole subject of bees and beekeeping is vast and there are a huge number of specialisations and learning opportunities within the overall theme to the extent that some enthusiasts don't even keep bees at all! One lawyer who was devoted to the study of bees, never actually kept any kept bees because he was fatally allergic to their stings, but he was unable to ignore the fascination of the subject and gave talks and lectures on the legal responsibilities of keeping bees in urban and rural locations and even developed an interest in bee biology.

Giving you an insight into ecology and food production and the protection of nature

With the above in mind you will inevitably end up with a unique insight into local and global food production, pesticides, diseases, droughts and floods and related factors that affect bees and other pollinators. Protection of the natural environment becomes as much your concern as it does any other naturalist or green organisation. The future of your bees is at stake and you can't but help becoming involved in that! Much of this knowledge comes about as a result of your own experiences with your bees on a small scale, almost without you realising it. What other pastime brings this knowledge to you in such a practical setting? Some beekeepers make a study of this aspect of the natural world and dive off on yet another tangent of study or community activity, for example, some beekeepers register with their local councils as someone for the public to contact if they come across a swarm. This is a really interesting off-shoot and great for learning more about bees – and bees that turn out to be wasps which are also fascinating creatures.

Learning about the bee colony and the science of bees

Let's firstly take a look at what else bees produce apart from honey. Remember, that they are master chemists. Everything they need to build and maintain their home, they make. They make a home from wax that they synthesise in their bodies and it is a home of geometrical precision. In this amazing honeycomb the queen can lay eggs, the bees can store food and the nurse bees can raise the brood. They collect food from flowers for themselves and the brood and produce the basic brood food royal jelly from glands within their heads. They have developed a sting for defence against predators including heavy-handed beekeepers and with a range of pheromones they are able to coordinate all activities within the hive including increasing the defence forces when needed. The colony can be regarded as a single organism with the bees as the individual cells of various specialisations and study of the operation of a bee colony keeps scientists amazed at both its simplicity and its complexity. The more you delve into this, the more you discover. Many beekeepers make the science of beekeeping a lifelong topic for study and enjoyment.

Discovering related insects

Most beekeepers fully realise that there are a myriad of other pollinating insects out there including the bumblebees and a wide variety of solitary bees. Many of these insects pollinate certain plants more effectively than honey bees and are used in some areas for that purpose. The alfalfa leaf cutter bee for example is a better pollinator of alfalfa (lucernes) than honey bees. Bumblebee packages are used for tomato pollination in green houses where honey bees don't prosper and where their buzz pollination technique does a better job than the honey bee. Butterflies and moths are also good examples of pollinating insects. The welfare of these insects is very closely related to that of the honey bee and their interrelatedness in nature is a very worthwhile topic of study and interest.

Evolutionary studies

Like man, bees provide an ideal window of study on evolution. Bees trapped in amber from over 100 million years ago show us in glorious detail how they and flowers evolved together in one of the most significant examples of an evolutionary symbiotic relationship that we know about today – a relationship that literally colours and feeds our world.

Microscopy

This can be a real eye opener and can become an absorbing pastime leading to a greater understanding of the physiological make up of this amazing insect. From preparing slides to studying the various features of the bee in incredible close up, microscopy offers many rewards to enthusiasts. With the convenience and versatility of modern microscopes and the ability to link them to a pc, you can easily take, keep and send close-up photos. It's not just bees either, some enthusiasts study – and marvel at – pollens and plant structures under the microscope.

Why not try photography

Another fascinating area is bee and insect photography. I have seen amazing (winning) photos by beekeepers which don't even contain a single bee but convey something magical about beekeeping, like jars of golden honey or multi-coloured hives in a stunning winter garden. From microscopic to landscape, there are so many visual treats to capture and display. There are many competitions available in most countries and some of these are devoted to beekeeping subjects.

Bee breeding

Some beekeepers devote their lives to breeding the perfect bee. Brother Adam was a prime example and the work of scientists both amateur and professional goes on. It can be an utterly fascinating project to breed a bee that is resistant to disease, gentle and a good honey producer. Many researchers stress that the development of this 'perfect bee' is imminent. We can only watch and see – or take part in it with our own research programmes at a local level.

Hive building and carpentry

For those beekeepers who have an eye for carpentry, this aspect of the hobby can be very enjoyable and can keep your costs down. It is astonishing just how many different beehive designs have been 'invented' and tried by hobby beekeepers engaged in hive design and building. Also, the number of large and small design changes that have been tried is unbelievable. Inspection panels, glass windows, different types of frame spacers, lifting devices and a million other gadgets have been developed and tried – some with great success and others with less.

Apitherapy? The medicine of hive products – huge!

This is a huge and well developed offshoot of beekeeping using products from the hive to treat human ailments. There is a wealth of anecdotal evidence to suggest that many of these treatments work and more recently some of these 'stories' have been proved clinically correct. For example, clinical trials have found that buckwheat honey has a more beneficial effect on coughing than most proprietary brands of cough medicine. Similarly the antibiotic character of New Zealand manuka honey has been clinically proven to be effective against the bacteria that cause stomach ulcers and the MRSA superbug, and is now used for wound and burn dressings by the UK and US health systems. Medical attention is now turning towards propolis and bee venom with clinical trials into their benefits to health continuing. It is not within the scope of this book to describe this important and complex subject in detail, but there are many well accredited courses where you can learn to become an apitherapist and many beekeepers and non-beekeepers are interested in this subject. Take a look at the Appendices of this book and see which courses can help you. For most beekeepers though, using honey and sprinkling pollen on their breakfast cereal provides a wonderful health boost.

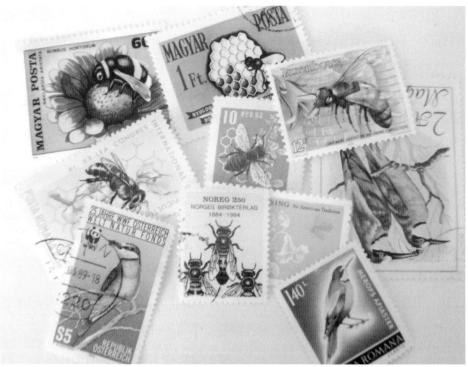

Stamps and postcards are a favourite with philatelists.

Bee philately

Stamp collecting is another interesting subject with enthusiasts from all over the world engaged in it. Bee eaters of all kinds, hive pictures and so on grace stamps from all over the world and collecting them is the subject of many websites, specialist magazines and books on the subject. Take a look at the website in the Appendices of this book.

Books, art and history

Just as stamps excite the imagination of many collectors, so do books, ancient and modern. With rare tomes on beekeeping going back centuries, it is especially exciting not only to collect these books but to learn what our forefathers knew about the natural world in their day and about bees in particular. The ancient Maya believed them to be messengers of the gods and even gods themselves and devoted entire religious ceremonies to them. The development of modern methods came about often with great resistance from the more conservative beekeepers of the various ages right up to modern times and the story is fascinating.

Collecting bee books is another offshoot of beekeeping.

For those who enjoy artistic passions, bees have always been a favourite subject for poets – just read the marvellous and thought provoking poetry of Sylvia Plath and Emily Dickinson, for example. Bees also exist in art, literature, music and ancient mythology and in some ancient cultures they were considered gods.

Cookery and brewing with honey

This is a very popular pastime and figures large in annual beekeeping association competitions and in the home. Making good honey cakes or a fine mead or honey beer is very satisfying if it's from your own honey. There is a definite art to making a good honey beer or mead and I confess that I am not the best exponent of this art – although not for want of trying, but I do love using honey in cookery and experimenting with honeys of different tastes.

Beekeeping! Not just a hobby but a goldmine of possibilities and a pastime of endless opportunities

In an age when it is evident that many people sometimes struggle to know how to occupy their spare time in a worthwhile and satisfying way, I hope now that you can see that beekeeping with all of its offshoots and options for study, application and enjoyment presents an absolute mine of possibilities for you. It can be a pastime that expands, or not, to occupy as much or as little time as you want to devote to it.

Warning

Beware however. The famous UK commercial beekeeper R.O.B Manley once said: 'If you're going to keep a lot of bees you're going to get bee fever, a form of insanity you'll never recover from.'

Are you tempted to explore further? If so read on!

2
Frequently Asked Questions about Beekeeping

A colony of bees will fly 90,000 miles, (about three orbits around the earth) to collect 1 kg of honey.

If you are still interested in beekeeping after reading the first chapter, you may now be at the stage of asking what you should be looking at next. What questions should I ask now? Who do I turn to now? Where would I keep bees anyway? What about my neighbours? It's well worth looking more closely into the various matters you need to consider before you commit yourself to beekeeping. This chapter aims to answer those questions so as to help you decide whether beekeeping is for you.

Here are some frequently asked questions for you to consider and the first question most people ask is usually concerned with stings, so it is worth getting this subject out of the way first.

Bee stings?

Everyone wants to know about these. I don't know a beekeeper who hasn't been stung at one point or another and if you take up beekeeping you will at some time undoubtedly be stung. If bees didn't sting, everyone would probably be a beekeeper! Bee stings can be very painful for some beekeepers and the effects of a sting may last for a while, but for many they are not that painful and the pain doesn't last for long at all. The pain varies from being a sharp irritation and then nothing more except maybe some itching and local swelling, right through to being very painful and often accompanied by a lot of swelling. However, recognising that they are a fact of beekeeping life, I don't know of any beekeepers who are especially worried about them. In fact they are rarely discussed amongst beekeepers and of little consequence generally. I know of no beekeeper that has been killed by a bee sting. Whatever other books may say, there is no cure for the pain other than using an anaesthetic. Onion juice, honey, vinegar, snake oil and other remedies are often touted but all are useless although perhaps of psychological value for small children. I have tried them all and have even tried scorpion oil which might make you live forever but it won't stop the initial sharp pain of a bee sting. With all of them, by the time the pain goes, it would have gone anyway. Bee venom is complex and clever and it is worth you reading more about it.

More serious bee stings

That is not to say however that there is no danger in this because in very rare cases there can be. In a tiny minority of people bee stings can be fatal. Just one will do. Anaphylactic shock can and does kill people and even experienced beekeepers who have had many stings may one day receive a sting that will sensitise them to venom and the next sting could be fatal. So what to do about this? Most beekeepers don't bother to do anything but the real solution is to always carry an injection of adrenalin around. These come in the form of an auto injecting EpiPen (so you don't have to actually shove a needle into yourself). The EpiPen® and EpiPen Jr Auto-Injectors (0.3 and 0.15 mg epinephrine) are used for the injection of epinephrine, the first-line treatment for anaphylaxis. The Auto-Injector is used to treat signs and symptoms of an allergic emergency, some of which include hives, redness of the skin, tightness in the throat, breathing problems and/or a decrease in blood pressure. Epine-phrine is a prescription medicine and so your doctor will be able to tell you if you are going to be okay with it and indeed must prescribe it. Allergic emergencies can be caused by other triggers in life as well such as food, medicines, latex, or even exercise. It must be emphasised though that this reaction is extremely rare and needn't put anyone off beekeeping. Statistics from the USA in 2000 (from the World Health Organisation) show that there were only 54 deaths attributable to bee stings – from a population of 281 million (Census data).

Now let's put that in perspective:

- In the same year, there were 15,517 murders in the USA (FBI crime figures).
- More than 20,000 people in the USA die from flu every year (US Centres for Disease Control).
- Even lightning kills more people than bee stings! On average, 90 people are killed every year in the U.S. by lightning. [NOAA Technical Memorandum NWS SR-193]

More recent statistics from 2010 in the USA show the following ratios:

Deaths from:

Car accident:	1 in 6,700
Murder:	1 in 18,000
On-the-job accident:	1 in 48,000
Bathtub drowning:	1 in 840,000
Lightning:	1 in 3 million
Hornet, wasp or bee sting:	**1 in 6.1 million**
Shark attack:	1 in 280 million

It is true that statistics can at times distort plain facts and admittedly, beekeepers will have a greater risk of being stung than members of the public but not a greater chance of a fatal reaction, so I hope that the statistics have reassured you about the danger of bee venom. There is little chance of you suffering from fatal venom allergy and if you take precautions such as carrying an EpiPen around with you and wearing normal protective clothing you will minimise or eliminate even these small risks.

Is it worth signing up for a beekeeping course?

Signing up for a beekeeping course at your local Beekeepers' Association (BKA) is strongly recommended by the National Bee Unit (NBU) a Government agency within the Department for the Environment, Food and Rural Affairs (DEFRA), which likens beekeeping to the rearing of livestock. When I started beekeeping I found the help and advice given to me by the local association in Lincoln to be more than useful for my beekeeping and I'm not sure that I would have stayed the course without their input and support.

Courses cost between £50 and £250 depending on their length, with winter courses generally focusing on theory before practical lessons are taken at an apiary in the spring. By taking a course you will be better able to decide whether to take up beekeeping. If you don't like it or have some fears, then you can save a lot of money by not becoming a beekeeper or by getting into a different aspect of beekeeping.

A course will help prospective beekeepers get a feel for the BKA and its members, their future colleagues. You will also begin to see how the BKA works to (hopefully) encourage and support new beekeepers and you will get to know fellow pupils many of whom will become your beekeeping buddies for years to come.

What equipment will I need and how much will it cost?

To the uninitiated, the equipment needed for beekeeping can look complicated but it really isn't and to get started you don't need much – a complete hive, a colony of bees, a smoker, a hive tool, an EpiPen and adequate clothing and that is all. There is a lot of equipment that you might want but don't need. Buying it later on to experiment with is all part of any hobby.

Beehives

There are lots of different beehives available, but really only two major types – moveable frame beehives and top bar hives of one sort or another. We will look into these different types in much more detail later. But how do you decide which is the best for you and for your area? One good tip is to find out what

Two WBC hives. My first hives and personal favourites for looks.

the beekeepers in your local beekeeping association are using and perhaps follow suit. If you use the same as them then they will be able to tell you best how to use your hive and all the various parts will be interchangeable with those of other beekeepers in case you need to exchange something or borrow something. Swapability is important and once you have a hive, generally the parts are not interchangeable with those of other designs of hive, so you are best to pick one type of beehive and stick with it. Choosing a hive type to work with deserves some consideration and the various advantages and disadvantages of the different types of hive will be looked at in detail.

Hives are probably the most expensive single item that you would invest in and you can buy complete kits from beekeeping supply companies. As well as hives, these kits include all of the basic tools and clothing, including a small honey processing kit for your harvest for around (at 2012 prices) £600, but if you intend to use the honey harvesting equipment of the local association you could then set up for around £400 or so. These kits are not bad value for money and at least you know that you have everything you need and that it is new and disease free. There are ways to reduce this initial expenditure such as buying beehives as flatpacks and assembling them yourself or even making your own hives from readily available plans on the internet.

Bees – if you want to buy your bees from a reputable dealer in what is called a nucleus box (or 'nuc') it will cost you anywhere from £120 upwards. For this you will get five frames of bees including a queen, stores and brood (young). Nuc's change hands far more cheaply within BKA's (often £50 to £80), and you also know where and who they come from and you can have someone check them out for you. Swarms are free but need a bit more planning. We'll look a bit later at the best strain of bees to start with.

Other items of equipment

Other equipment you will need are:

The smoker – essential for calming the bees as you inspect them. The theories behind this statement are many but smoke probably confuses the bees' senses in some manner and calms them. Most advice suggests getting a big smoker with stout bellows. For many, this is good advice. Small ones run out of fuel quickly and need constant relighting but new types of fuel (compressed cotton waste, for example) and the fact that you only have one or two beehives means that a smaller, less cumbersome smoker could fit the bill. The cost of a large smoker will be around £30–£50 or you can even buy a clockwork one for around £80!

A good smoker is essential.

The hive tool – simply a piece of metal usually of one of two designs. This most essential piece of kit is used for forcing hive boxes apart, separating and lifting frames from hives, scraping wax and propolis from hives and a million other tasks that need a bit of brute force. Any home metal worker can make these but from a bee supply company they will cost around £10.

Hive tools are easily lost so keep spares.

Good clothing gives you peace of mind.

Adequate clothing – economise on other items but not on this. A good bee suit will protect you from stings and last for years. These can be purchased from around £60–£125 and a good pair of gloves from around £15. Some beekeepers don't use gloves and put up with the stings. That isn't for me as I don't find it at all enjoyable so I use gloves. Some beekeepers swear by rubber washing up gloves which take away the worst of a sting and are very good for 'feel', and if they lasted longer than a few minutes, I would use them more often. However my real preference is for kid leather gloves with attached gauntlets. I find that these soft gloves give me a decent 'feel' for what I am doing while protecting me from stings at the same time while the gauntlets do a good job of protecting against crawling bees.

Running costs (based on two hives)

There are running costs in any hobby or pastime and beekeeping is no exception. These are very reasonable and the following should be included:

The EpiPen or adrenaline injections – very much advised if the doctor okays it. The adrenaline injections are very much cheaper than the EpiPen but you have to inject yourself which many find difficulty in achieving. An EpiPen costs around £30–£45 and should be replaced before its expiry date, so it's definitely part of the annual running costs outlined below.

Varroa treatments – these will include annual treatment costs for varroa mite infestations – currently around £50 per year for a couple of hives but very much cheaper through your BKA.

Membership of the local beekeeping association – will also be a running cost. It's not compulsory but highly recommended and includes insurance so it's well worth it. Typical benefits include:

- Membership of the local association, which is usually part of the British Beekeepers' Association (BBKA) and an immediate introduction to beekeepers of all ages and competencies.
- Third party, public liability and bee disease insurance.
- Often included is 'BeeCraft' – the official monthly journal of the BBKA.
- The British Beekeepers' Association Newsletter.
- The local association monthly newsletter giving details of future events, topical information and an opportunity to exchange news. Most of these newsletters have sales and wants sections where you can buy equipment and bees at secondhand prices from known sources.
- Contact with other beekeepers and associations.
- Beekeeping equipment at reduced prices.
- Opportunity to participate in the association annual Honey Show, the local County Show and the National Honey show.

Typical costs £45–£50 – one of the best deals in the UK, the local association is your front-line support group and will be essential in your early years and after.

Total costs

So around £700 max, plus or minus a few running costs listed above, will get you fully equipped and ready to go (much less if you make your own hives or buy a cheaper type of hive). Just like any other hobby however, there are a million add-ons that you can purchase and marching around a bee supply supermarket with a trolley will often see even seasoned beekeepers piling up loads of junk and gadgets both useful and useless that they 'intend' to use but often don't. Compare these costs with other hobbies – fishing, golf, cycling, stock car racing, aero-modelling and so on and you'll find it all very reasonable. And, you get honey out of it and a lifetime of learning! Using secondhand hives and 'free' swarms of bees can reduce this initial investment and a later chapter will look at the advisability of this.

What are the health requirements?

The ability to lift boxes with honey in them is of course a requirement for most beekeepers and these boxes can be heavy. A Langstroth deep box full of honey can weigh around 30 kg plus – one reason for taking care in your choice of hive. However there are many ways to mitigate this: using hive barrows and lifting handles; using smaller hive boxes for the honey storage areas; using the help of other people in lifting and so on. By placing hives on hive stands beekeepers can avoid too much bending when inspecting their hives and I do know of a disabled beekeeper in the UK who keeps bees using a wheelchair to visit his hives, and a commercial beekeeper who earns his keep from bees in New Zealand who is totally blind!

Top bar hives such as the Kenya Top bar hive require much less heavy work and for those people who are unable to lift heavyweights, these are a very good idea. These hives (and others) are discussed in detail later in the book.

How much time and effort is required?

One benefit of beekeeping is that you can spend as much or as little time on it as you want. Obviously there is a minimum requirement to care for your livestock and this duty of care should not be taken lightly, but as long as you keep up a fairly regular pattern of inspections (mainly for disease or other problems) and ensure that your bees are adequately housed, have sufficient stores of food and sufficient room, you can leave them alone for the rest of the time. This means that there is no need for you to 'be there' all of the time. If you need to go away for a few weeks on holiday, just ensure that your beehives are secure and leave the bees to it!

Where could I keep my bees?

New or prospective beekeepers should not be over worried about where to keep bees. Bees are not only kept in remote places in the countryside and in fact there is an increasing trend toward urban and especially suburban beekeeping. The most important question to ask about where to keep bees is, 'is there enough forage for them?'

This question of placing beehives is probably easier to answer than you would think, even if you live in town. Bees are hardy creatures and can live in a variety of situations but obviously there are sites that are more suitable than others and there are some sites that are totally unsuitable. However, whether you live in the countryside or in the city, you can usually keep bees and safely locate your hives. Urban areas really are increasingly becoming a favourite location especially in cities.

According to the London Beekeepers' Association 'urban' bees have a wide range of forage, as the gardens and green spaces in cities contain a rich variety of trees and flowers. This, and the slightly milder weather, means that the beekeeping season is longer and usually more productive than in rural areas. Gardens cover over three million hectares in the UK which is more than all the nature reserves put together and they offer continuous forage from March right through to September. You just don't often get that period in the countryside unless you move your bees from crop to crop.

Increasing numbers of urban beekeepers are taking to the rooftops and small gardens of the towns and cities of the UK, Europe and the USA including some of the more exotic names around. The London Stock Exchange (LSE) has installed apiaries on its premises. The CEO of the Exchange is a keen beekeeper and hopes to involve local groups in his initiative.

Other urban beekeeping organisations include the Tate Modern, the Bank of England, St Pancras station and Fortnum and Masons. This last firm has installed its hives on the roof and the bees have full access to the gardens of Mayfair, one of the best areas of London. Superior honey from superior gardens! What better advertising! With the installation of two hives on the roof of the LSE, these landmarks will all have their own bee colonies – part of a trend towards eco-awareness in the city.

The website http://urbanbees.co.uk may be a useful reference for you if you are considering joining this rapidly growing movement.

Siting urban and suburban bees

Here are the things to think about when siting urban bees:

- **Neighbours.**
  Neighbours may like to read about and sympathise with the plight of bees and the ecology but may not like the idea of actually confronting a bee for real, and certainly not a swarm of them! Biodiversity, ecology, nature and so on are fine words unless they sting you.

A typical suburban apiary.

- **Rooftops are ideal**.
  Just make sure that access is not too difficult especially if you need to carry heavy boxes up and down steps.
- **Small gardens with high fences and walls are ideal.**
  Bees fly upwards to get over the fences and so are above neighbour head height. If there is no high fence in front of your hives, place one there or for the longer term plant a screening type of bush/hedge.
- **Try and obtain a gentle strain of bees.**
  Every bee seller will claim their bees are gentle and super productive but just be careful and obtain some good advice from the local beekeeping association.
- **A water source is essential.**
  A real must otherwise your bees will use someone else's water source and this can cause problems. Some of my bees in Spain infested my neighbour's goat troughs and his goats were not happy about it. It caused me no end of trouble sorting that one out. If you have a garden pond or live near a natural water source such as a river or lake, that is ideal, but if not, how about a dripping outside tap. This is an easy water source for bees especially if you let the small flow of water run through a small bed of pebbles or stones.

- **Insurance.**
  Insure your bees for third party liability. If you join the local beekeepers' association, this will be provided as part of your fees for a certain number of hives.

Rural locations

Rural locations are often more difficult to find than urban sites. The following points are worth thinking about:

- Try and ensure that anywhere you place bees has easy access preferably by vehicle. I was advised of a supposedly 'perfect' site in Spain by my new neighbours who even offered to help me move my bees there. After scrambling down a sandy and very steep slope with the hives coming apart and my neighbours blasting the attacking bees with flames from the smokers, we finally came across a cliff face. Lifting hives above our heads, we finally got two of them onto a plateau on top of a small mountain, and exhausted and in pain, we sat down to rest. I told my neighbours that it was the worst possible place in the entire world to put bees. He answered, 'perhaps senor. But look, just look at the view they will have!' He had a point. He said it just as the dawn broke and the sun rose at our backs and shone on the beautiful Mediterranean sea view below. It was still the worst place in the world for bees however and we still had to get down.
- Make sure your bees have nectar sources within less than a mile of your hives, preferably nearer. It is easy to imagine that if you are in the countryside there will be flowers but that is not always true. A thousand acres of wheat gives you toast but no honey. Bees not only need nectar but also need early sources of pollen for a good colony build up. This is a requirement that many beekeepers – even experienced ones – forget about and ascribe the resulting poor colony build up to other factors. Your association will be able to guide you on this point.
- Just as with urban sites, remember that bees need a water source. This is unlikely to be a problem in most parts of the UK or Northern Europe but remember my goat story above and ensure that there is one somewhere.
- Placing hives under trees has its good and bad points. True, bees can take up residence in tree trunks in the wild but they won't be dripped upon inside a tree. They will be if a hive is under a tree canopy and they don't like

A rural apiary. Make sure there is plenty of forage.

this. On the other hand, especially in hot areas, the dappled shade from trees offers a very good method of helping the bees maintain hive temperatures at the optimum levels.

• Make absolutely sure that your site is not prone to flooding in the winter and is not a frost hollow and always place your hives on hive stands to raise them off the ground. This ensures that the hive interior doesn't become damp and allows air to freely circulate under and around the hive.

• Keep beehives as much as possible out of site of nearby roads and paths so that your bees are not vandalised or stolen and obtain insurance to cover these eventualities.

• If cows or sheep are going to share the ground, make sure that the hives are protected by a temporary electric fence. Cows use hives as scratching posts and are generally clumsy creatures and always seem to knock unprotected hives over. Horses can panic near bees and in my view it is best to keep them well separated.

How will children, neighbours and pets be affected?

Beekeeping is an ideal hobby and educational experience for children and you could always encourage them to join you. However, if that fails, talk to them about your bees and let them know that bees don't sting people for the fun of it and if they leave the bees and the hives alone there will be little problem. Keep them informed about when you will be inspecting the bees and when you will be taking off the honey and any other times you will be attending to your bees, and with common sense there will be minimum risk. The same advice applies to neighbours as well. It is worth repeating here my earlier statement that neighbours may like to read about and sympathise with the plight of bees and the ecology but may not like the idea of actually confronting a bee for real, and certainly not a swarm of them! Biodiversity, ecology, nature and so on are fine words unless they sting you. Tell them about your bees, give them honey and invite them to the odd association function. Let them know what's going on but do be aware that any sting even from a stray wasp will be blamed on your bees and so education is your best defence. Another possible source of complaint from neighbours will be from bees soiling your neighbours washing when they fly from the hive and excrete. This is difficult to stop but the area covered is generally along the flight line to the hive and so you could try changing the direction of the hive entrance. The siting of the hive has to be considered carefully to minimise such problems. Pets can be a different matter but after one sting, most will keep well away.

Do I need permission to keep bees, even in my garden?

In the UK there is no legislation to stop you; this may not be the case in other countries. Unless you have a larger garden it would be advisable to keep no more than two colonies. Remember that to avoid problems you should use colonies that are gentle in nature and this may mean obtaining good advice from the local beekeeping association. Don't be afraid to call for experienced help if there are any problems.

Although there are no laws as such mentioning keeping bees in the UK, prosecutions against beekeepers have been made on the grounds of 'Statutory

Nuisance'. 'Any animal kept in a place or manner which is harmful to health or a nuisance', 'Whereby interference with the enjoyment of the neighbouring property is a prerequisite'.

There may be separate rules for keeping bees on an allotment and you should always seek advice on this with the allotment secretary. Even the greenest of gardeners can put their collective feet down when faced with a truly green situation. Remember, laws change regularly, especially council bylaws and it is always wise to keep up with them. This is another good reason for joining the local beekeepers' association because they will be able to advise you on all issues affecting members.

Is insurance required?

Yes, but not by law. Your association membership will usually include third party insurance and product insurance for up to a certain number of hives. If you live out in the countryside and keep bees in more remote places then theft and damage insurance should be added. You never know when a forest fire is going to sweep through the area – which is how most of my bees were destroyed in Spain, so it can happen.

What support will I get and from whom?

Your primary support group will be your local branch of the BBKA or equivalent home country or Irish associations and as I have said before, I would strongly advise joining up. It is a good idea to visit one well before you decide to start beekeeping and I'll explain why in Chapter 3. This organisation will be your support group. It will often appoint a mentor for new beekeepers and will keep you on the straight and narrow. It is in no ones interest to have a bad beekeeper in the area. Don't even think about it. Join it.

The object of the BBKA and its local branches is to promote and further the craft of beekeeping and to advance the education of the public in the importance of bees in the environment.

From its informative website it states that:

Being a member of the BBKA gives you these benefits:

- BBKA News
- Public Liability Insurance
- Product Liability Insurance

- Bee Disease Insurance available
- Free Information Leaflets to Download
- Members Password Protected Area and Discussion Forum
- Correspondence Courses
- Examination and Assessment Programme
- Telephone Information
- Research Support
- Legal advice
- Representation and lobbying of Government, EU and official bodies

The Welsh (Cymdeithas Gwenynwyr Cymru), Ulster and Scottish as well as the Irish Associations offer similar benefits and if you are in these areas then again you should approach them for advice in the first instance. See the list at the back of the book for contact details.

In the UK, FERA (the Food and Environment Research Agency) holds under its wing the National Bee Unit which in turn controls a set of regional bee inspectors. These inspectors are beekeepers of great experience and are trained to answer just about any question you care to ask – and are most willing to do so. They will visit you if required and will often be seen giving talks to the local beekeeping associations.

Reading

There are also many very good books on the subject of beekeeping for you to read and refer to before you take up beekeeping and most can be found in or ordered from libraries. The local beekeeping association will probably have a small library as well. Books offer advice and guidance on all aspects of beekeeping and will probably be your first port of call if you take up beekeeping. Websites offer an almost unlimited amount of material and many of the most useful for you are detailed at the end of this chapter.

There are probably many other questions that you may have about beekeeping to help you decide whether to take it up but I hope that the information given above may have given you some food for thought and at least answered the most important questions you may have. If you are still interested, Chapter 3 will look more at some of the practical problems associated with keeping bees and how to overcome them.

Still interested? Keep reading.

3

Hives

*The bee's brain is oval in shape and only about the size of a
sesame seed, yet it has remarkable capacity to learn and remember
things and is able to make complex calculations on distance travelled
from the hive and foraging efficiency.*

Once you have read this chapter and perhaps spoken to other beekeepers in
the local association, you will be better able to make an informed decision
about the hive type that you will adopt if you decide to take up beekeeping.

So how many hives should I start with?

I advise that you start with two hives. Using one hive can present problems.
Bees do suffer from a multitude of ailments and other problems and if your
one hive crashes, e.g. the queen dies, then having another hive to draw bees
and brood from is an important consideration. It can keep your beekeeping
hobby going. The mechanics of all this will be explained later in the book but
suffice to say here that more than one hive is a good idea to start with. You
may ask, why not three, or four or ten hives? The answer to this is that when
starting any hobby, you can initially be overwhelmed – especially when dealing
with live charges and there are a lot of live charges in a beehive – around 50,000
in summer! So initially it's a good idea to keep everything to manageable
proportions. Just make sure that you can deal with two hives before increasing
numbers. This doesn't just apply to managing the hives, but also dealing with
swarming, diseases and other problems, and of course harvesting your honey.
Taking and extracting honey from say 15 hives is a lengthy business unless you
have very good and expensive equipment, especially when you are trying to do
all this in your kitchen, and the number of bee boxes you will need can be
daunting. I can assure you that it is very easy to increase the number of colonies
you own in year two. Almost too easy in fact and many beekeepers have
difficulty keeping the numbers down.

What type of hives should I use?

I didn't have to think about this question when I started because I wanted WBC hives. I saw a picture of them and they looked like bee hives ought to look. Nowadays, however, there is a lot of opinion out there and you may wish to look at all your options in a more considered way than me. My personal advice would be to use the most common moveable frame hive in your area but there are alternatives to this and it is important that you start off on the right foot with the right hive for you. More hive types are now coming on the markets including those that are believed to be 'more natural' for bees – and I certainly don't sneer at these alternatives. They look interesting and have a lot going for them. For the hobbyist with just a couple or so hives, I don't think it matters much which hive you choose as long as you look after your bees.

When you have read this section and decide that you want to start beekeeping, have a word with some local beekeepers in the association and see what they have to say as well. You may find that many beekeepers will dismiss any type of hive that they personally don't use but some will be more open minded and it is worth finding someone to discuss alternatives with, especially the recent trend towards top bar type hives.

Let's now take a closer look at the two basic hive types and see where we go from there. It is worth some thought because once you have your hives you are stuck with them unless you expend yet more money on changing everything. I am sure you will read and hear conflicting ideas and advice and so I have included discussion about the various points involved.

Hive types

There are two basic types of hive; those with removable frames and those which use top bars (which are actually removable as well). Both types have their advantages and disadvantages. For the commercial operators that rely on fast and efficient movement of hives from crop to crop for pollination or honey, it is best to rely on one type, whereas the hobby beekeeper is able to pick and choose from any.

Removable frame hives

These hives are very simply boxes stacked on a floor and capped by a waterproof lid. Inside the boxes hang eight to ten rectangular frames in each box and each frame holds a honeycomb or sheets of foundation. Foundation

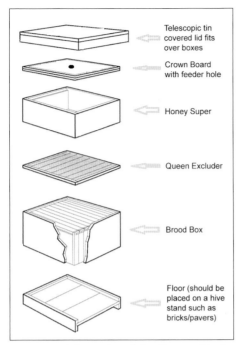

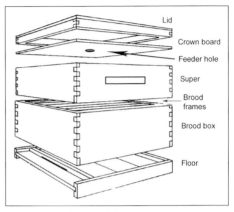

Telescopic tin covered lid fits over boxes

Crown Board with feeder hole

Honey Super

Queen Excluder

Brood Box

Floor (should be placed on a hive stand such as bricks/pavers)

Lid

Crown board

Feeder hole

Super

Brood frames

Brood box

Floor

Figure 1 and 1a: Removable frame – Langstroth type hive layouts.

A typical removable frame hive.

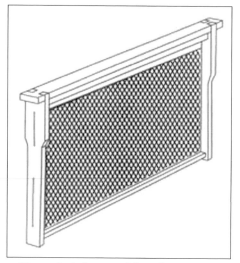

Figure 2: A frame of foundation wax.

sheets are thin sheets of beeswax cut to the rectangular shape of the frame and stamped with the hexagonal imprint of honey cells. The beekeeper fits them into the frame and the bees 'draw out' these imprints using their own wax and so produce honeycomb. This foundation can be strengthened with wire melted into the wax which helps keep the honeycomb in one piece when the beekeeper removes the honey in an extractor.

Most people will have seen pictures of honeycomb in bees' nests and you will notice that they hang downwards and are spaced apart so that the bees can crawl between the combs and so use both sides for storing honey.

This is replicated in a man-made beehive by using rectangular frames hanging in a box from a revetment and that all important space between the individual frames and the sides and roof of the box is also maintained. If the space is less than approximately 9.5 mm (3/8 in) the bees would tend to fill it with wax. If it is larger, then they build brace comb in the gap to help anchor the combs to the sides of the cavity or the roof. Discovering this simple fact took beekeepers centuries and it wasn't until the 1850s that the Rev. L. Langstroth in the USA became the first beekeeper to use this 'gap' knowledge in a beehive as what he described as 'bee space'. Because of it he was able to remove inspect and replace each frame of honeycomb without tearing it away from the hive side wall or from other frames and so destroying the nest or honey storage area. Most beekeepers that you will come across use this type of hive. Using these removable frames greatly

Wild comb showing how bees keep a space between the combs.

facilitates hive inspections which are important not least in ensuring a disease free colony.

The hives themselves come in different sizes but in the UK, most will use hives of the National size or Smith (Scotland) size. Commercial beekeepers tend to use slightly larger hives based on the Langstroth model or the Dadant model. The Langstroth is probably the most used around the world. The differences in size need not unduly worry us at the moment because I would advise the hobby beekeeper in the UK to stick to the National or Smith size of hive unless you know now that you want at some time in the future to build up to commercial status. One hive that sets itself apart from all the others is the WBC named after William Broughton Carr but only in that the usual rectangular boxes are surrounded by an outer cover which makes the whole thing look like a proper traditional beehive.

The inner boxes are usually of British National size. If I was going to start beekeeping all over again with two hives, I would get a couple of WBCs with a National size set of inner boxes. In fact I did just that years ago. These hives look really good especially if painted white and for the hobbyist are practical and 'satisfying' to be around. Because of the extra parts however, they are more expensive if purchased new. For those beekeepers skilled in carpentry, free plans are available on the internet for most hives and a list of useful websites are visited in the Appendices of this book.

A WBC inner – similar to any other removable frame hive. National size.

Another more unusual looking hive has recently arrived on the market. It's called the Omlet hive and is made of plastic and more importantly it uses a mesh floor which I believe to be an important feature of any hive design. It closely resembles those letter boxes seen on American TV films that sit at the end of the driveway. It is an ingenious design again based on National sized frames and comes complete with everything needed to get going. Check it out on its website – www.omlet.co.uk and read the comments of users on the forum to see if it would suit you.

The beehives that I have mentioned are the most common beehives to be used in the UK and if you use one at least you will have support from other beekeepers using the same type of hives. You will also easily be able to obtain spare frames and other parts at a reasonable price.

Floors

Beehive floors are either made of solid wood or stainless steel mesh. I firmly believe that mesh floors are better by far for the bees. Some beekeepers believe that in cooler areas mesh floors would make the hive too cold and draughty for bees. But bees don't heat up their hive, they keep the brood nest at an even

An Omlet hive – based on National hive frame sizes.
An ingenious and useful urban design.

temperature and that's all they need to do. (Bees nesting in tree hollows don't have floors.) Mesh floors have the added advantages of preventing water build up in hives after rain; they provide for better ventilation of the hive; they allow varroa mites to fall out of the hive and so aid in varroa control; and they prevent a build up of debris in the hive. There are so many advantages for the bees that it would be foolish not to use them. I recognise that many hives arrive with solid floors as the norm but over time, see if you can change to mesh floors. They are better by far.

Mesh floors are an aid to keeping your bees healthy.

A Kenya top bar hive.
Notice the sloping sides to prevent comb sticking to the side walls.

Top bar type hives

What is a top bar bee hive (also known as the Kenya top bar hive)?
These hives were really designed for use in developing countries because of their cheapness to manufacture (DIY) and ease of use. They were first designed in Kenya hence their name, although variants were to be found in many cultures and in Western countries are often considered by some to be perfect solutions for back garden beekeepers. They can be made from salvaged materials, and even hollowed out logs.

The Top Bar Bee Hive is so named because there is no frame (no side or bottom bars), just a 'top bar'. Like the national or similar hives, top bar bee hives also have removable frames so you can collect the honey, but the frames are added along a horizontal plane instead of vertically building up with

Figure 3: A Kenya top bar hive.
The bars run horizontally rather then stacking boxes on top of each other.

stacking boxes. The bees build their comb to hang down from the individual bars (just as they do in the wild) and so the beekeeper does not provide a wax foundation in a frame for the bees.

The hive body is V-shaped in order to keep bees from attaching the comb to the hive-body walls, which would prevent the beekeeper from being able to lift the bar out of the hive.

One drawback of a top bar bee hive is that honey cannot be extracted by the usual honey extractors because a top-bar frame does not have reinforced foundation or a full frame with side and bottom bars. The combs have to be cut off and either left to drain or squeezed out. Or you could produce cut comb honey which is a delicious and really saleable product. Just buy a comb cutter, lay your combs out, cut them to shape and pop into a plastic container. The combs could also be placed in a cappings extractor and be spun out. This is a device used by beekeepers who wish to spin out the often considerable amounts of honey left after they have scraped off the wax cappings of a honeycomb to better spin out the honey in the cells. This means that you will get less honey but a lot of wax! And the bees have to rebuild the comb after each harvest, making the honey yield less than traditional hives. Bear this in mind when choosing a hive.

Top bar hives are especially valuable for those who don't want to or can't lift boxes heavy with honey for one reason or another. They will find these hives ideal as long as their disadvantages are kept in mind.

The Warre top bar hive

The Warre hive is also fairly simple in concept because even though it is a vertical stacking type hive, it also contains no frames to which the combs become attached, but simple 'top' bars just like the Kenya hive. Like the Langstroth hive, more boxes can be added as and when the colony builds more combs, the fresh box added with little disturbance to the bees.

It was designed by French beekeeper Abbe Warre and the design was supposed to be as close as possible to the bees' natural home. It is made up of a series of boxes (or supers) stacked on top of one another – just like a Langstroth or National hive.

The bees instinctively start building comb at the highest point (in the top super) and work their way down through the supers one by one. Happy bees in a safe and well nurtured environment are at the central philosophy behind the Warre concept. The concept also encourages minimal interference with few

A Warre hive.

inspections, but rather using observation at the hive entrance to determine the health of the bees. Observation windows are a newer feature of these hives and allow beekeepers to better see what is going on inside. Opening up the hive is believed to be detrimental to colony health because it releases heat and hive odour and disturbs the bees.

Learning to interpret what is going on in a hive by observing the hive entrance is a good idea for any beekeeper because you can then at any time when you are passing your hives on your way to work/pub/church etc. get a reasonable idea of what is happening inside. One basic example of this is that if the bees are flying freely but there is no movement from one of the hives, then you need to check it out and find out why. But as I found, you can't do this without opening it up.

There is a very good book on the subject of hive entrance examinations by H. Storch called *At the Hive Entrance* and it was one of the first beekeeping books that I read. I soon found, however, that although the information was useful and very educational, I was missing too much by not inspecting the bees inside the hive often enough. Maybe that was due to inexperience then, but now as a fairly experienced beekeeper although I can still see the merits of Storch's idea of inspecting by looking at the hive entrance, I have now summarised and in some cases updated all the points in the Beekeepers Field Guide which is a beekeeper's operations manual. I can say here and now though through considerable practice and experience that in this day and age, external inspections simply are not sufficient. They are useful but secondary. Glass windows used as inspection apertures are not good enough. They will not pick up American Foul Brood (AFB) or European Foul Brood (EFB) until it is too late. However well you look after your bees, they will contract AFB if a nearby neighbour's bees have it. Inspecting your hives internally is essential to ensure that you are able to at least try and prevent problems and diseases. Also, there is the point that beekeeping should be fun, educational, and instructive and beekeepers should be encouraged to open their hives at reasonable intervals and watch their bees at work inside the hive.

And your neighbouring beekeepers will not thank you if your bees have diseases and you have not properly observed the situation in your hives and so done nothing about the problem until it is too late. Diseases spread, so especially beware those who advocate (and some do) not carrying out hive inspections. You cannot see the beginning stages of some diseases for example without looking closely at the combs.

Do the bees care what type of hive they occupy?

I often think (and have read) that many proponents of alternative hives have taken various concepts about a variety of matters such as selection and genetics, hive heating, better hive dimensions, using recycled timber, natural comb building, swarming, quilts over the top, chemical-free beekeeping and so on, and have applied them to these hives as though the advantage was in the hive design itself without having researched the issues in a scientific way. I'm not sure that the bees care, so what practical experience do I have for saying that?

Science is thankfully returning (gradually) to a sort of understanding that we're all part of a living earth – almost a Gaia approach. I have believed this for years and so my bees in Spain were synthetic chemical free (I was a registered organic beekeeper) and I treated varroa with olive oil and thymol; I 'supered' below (can that be the correct term?); I used starter strips which are cut down pieces of wax foundation for most of my boxes (I couldn't afford anything else at the time) – I used hive entrance inspections and only opened up when I had to. My hives were exceptionally cheap and were made of pine from fully sustainable forestry plantations. They were treated with linseed oil for wood protection and the lids were painted white to help with cooling (definitely no quilts required). I removed early Spring honey and left on Autumn honey for the bees. My bees were fighting fit (literally) and very healthy without the use of harsh chemicals and produced excellent honey crops seemingly from nothing in the dry, arid heat. My bait hives attracted bee swarms that happily – or at least voluntarily (I assume) took up residence in them even though they are supposedly 'not natural' in size. In fact they were the type of hives that the natural purists now scoff at – Langstroth hives with standard frames and I 'farmed' bees commercially.

I do believe that hive choice is a matter of horses for courses and would not like a new beekeeper to be swayed too much by the word 'natural'. I'm not at all convinced that a bee colony would set up in a Langstroth bait hive if it wasn't in their little sesame seed sized minds 'natural'. I labour this point a bit just to make sure you understand that you don't *have* to have a Warre hive or a top bar hive to be an effective beekeeper. There are plenty of other reasons why top bar hives might be more appropriate for you as a beekeeper but I really don't think that the bees care as long as you look after them properly. I did have experience of keeping top bar hives (cork hives) as a hobby idea with natural comb hanging beautifully from the cross bars.

Requirement for proper hive research

Proponents of alternative hive types say that they have 'noticed' that bees kept in their hives and kept according to their more natural methods show much reduced disease and other problems. Proponents of organic beekeeping claim similar results (I did), but anecdotal evidence, however well disposed you are to it, can be misleading. Others point to the fact that even Langstroth himself thought that an increase in bee diseases had occurred after modern bee hives were developed. (This could be however partly because before these hives were developed, bees were often destroyed at the end of each season and the honey taken. In other words, a colony would last one season – hardly long enough for any bacteria or viruses to stand a sporting chance of building up.) If you kill the colony, you kill the accompanying mini beasts.

This isn't to say that the evidence, or the conclusions proponents derive from it, is incorrect, but it might be. There has been to my knowledge no proper scientific research to validate any of the claims and unfortunately, to carry out the research it would be necessary for there to be a large test sample and probably some funding over a decent period of time. Perhaps it is possible for groups and organisations to combine to do this and I believe it would be a well worthwhile exercise even if it took some years to accomplish. Bees and other insects are disappearing and if there is an answer to this in using these hives and methods, let's investigate it – properly and scientifically.

Warre or other top bar hive?

My answer would be that I wouldn't not recommend one. I am all for ensuring that bees are maintained to the best of anyone's ability in the most natural way possible. Hobby beekeepers are in the best position to pick and choose and they have time for experimentation with any hive type they want. As long as this is done sensibly in the bees' interest and with disease avoidance kept uppermost in people's minds (partly because that is the law of the land), then there is no problem. There is a list of websites in the Appendices of this book that may be of interest for those new beekeepers who choose top bar/Warre hives and also a list of books of interest. Use them and their advice and don't become isolated. You may have to go it alone in your area though.

Final advice: don't become isolated in your choice.

4

Bees

Honey bees are the only insects that produce food for human consumption apart from the Lac insect the excretion of which is used to coat some candies.

This chapter continues to look at the occupants of the hives and discusses how a colony of bees operates. I have been with many new beekeepers who are astonished and sometimes very much alarmed when I open up a hive so that they can see the bees inside, but this is an essential part of beekeeping and so in this chapter we take a look at:

- Going to the local beekeepers association for advice/information.
- Visiting an apiary and looking into a hive.
- Going to beekeeping classes to see if you would like to manage a hive or two.
- Learning about the bees themselves – a brief run through some basic biology: microscopy, mating, navigation and communication.
- Understanding the colony/how it works.
- Colony nest requirements.
- The role of the beekeeper.

Hopefully, the previous three chapters will have answered the majority of practical questions about beekeeping and you won't now be in the least worried about bee stings, killer swarms or angry neighbours and you'll be certain in your knowledge that statistically you are more likely to be killed by a horse falling on you than by a bee sting. You will have an idea of where you could keep your bees and what to keep them in and you will be more aware of the type of equipment that you would need and relative costs.

Seek local advice

The best thing to do now is to contact your local beekeeping association if you haven't already done so. Beekeeping associations are a vital resource. Not only can they help with your introduction to beekeeping and keep you up to date with regulations, rules and government policy, but they can also, if you want them to, be a source of social activity and friendship. The library or council websites will give you contact names and numbers or you can call them for further advice. The BBKA, Welsh BKA, Ulster BKA and Scottish BKA websites are mines of information. A complete list of useful websites is located in the Appendices of this book. By the way, despite its name, the BBKA only operates in England.

Your local BKA can offer winter courses and lectures on beekeeping, visits to other beekeepers' apiaries, classes, training in most aspects of beekeeping, will give advice on local regulations relating to town councils and allotments and so on, and will be able to advise you of the suitability of sites. Most BKA's have their own apiary and one of the options for siting a hive would be at the local BKA apiary. This may be a good idea as help is usually at hand and you can watch lots of more experienced beekeepers at work. You may have to wait for a place but beginners often take priority. You will be able to talk to local beekeepers and ask them what they think about any aspect of the hobby – they will all be very willing to talk to you. They will show you their hives and most will let you 'have a go' under their supervision and using a borrowed bee suit, hive tool and smoker. You will be able to look into hives full of bees and try and find out what they are all up to. Look at thousands of bees crawling around and have thousands of bees stare back at you! Not literally of course but you will be able to hold up a frame of bees and watch them go about their business. Actually most of the bees will ignore you and move about apparently at random – but it isn't really at random. Everything they do, they do for a purpose and with experience you will see this and begin to learn to read the hive. You may see the queen moving purposely about on the comb and if you do, quickly replace the frame into the hive before you drop the thing and lose someone's valuable queen. I did just this on my first apiary visit. The beekeeper was very nice about it all though.

But the best thing is that by doing it you will be better able to decide if you want to continue before you commit any money to the project. If at this stage, you don't feel that it is for you then you won't have wasted a penny – and you will have learned something at least about a remarkable insect, *and* you will have chatted to some very interesting people. Not bad at all. Also you will at least have an idea of really what it is all about and have a glimpse of an interesting future that lies in store.

The hive occupants

Choosing hives and deciding where to site them are one thing but sooner or later you will need to fill them up with bees!! You will actually have to obtain a swarm or nucleus or package of bees and introduce them in their many thousands to their new home. But how should you choose them? Where do you get them from? How do bees operate? What do they do? What are the types of bee that live in the hive? Do you need to feed them? How do they organise themselves? You will no doubt have dozens of questions but hopefully before you obtain your bees, you will have spoken to the local BKA members and maybe even seen a hive of bees at first hand. But if you haven't, don't worry, bees are okay and once you have a basic understanding you will be much happier about grasping the nettle and getting your bees. Firstly though, where do you get bees from and how do they arrive in your apiary?

Bees. These are your charges.

Where to get bees from

If you buy a couple of hives of bees then obviously you get the bees that live in the hives and that's that. The vendor may or may not know the strain of bee they are offering but should be able to let you know their characteristics: good laying queen, gentle bees, good honey producers, don't need much feeding and so on. Do obtain the guidance from another experienced beekeeper and do be careful here of gross exaggeration on the part of the vendor by taking along an experienced beekeeper. On one occasion concerning a sale to me, the extra gentle, super productive, virile bees that I thought I'd bought turned out to be savage little horrors that swarmed incessantly and produced honey by the thimbleful. I should have known better. I wouldn't dream of comparing beekeepers with used car salesmen and fishermen but just remember that the vendor will have looked after and loved their little darlings prior to sale – and simply now can't wait to get rid of them. To you! For money!

Starting with a swarm

Buying hives and setting up and then obtaining a swarm from a local beekeeper is another way of starting out in beekeeping and this is what I did. I purchased two WBC hives from the Lincoln BKA annual auction of bees and equipment, made up frames with foundation and placed them in the hive all upside down and added the bees. It took me ages to sort it all out. With a swarm, the provenance of the bees may be a mystery as will their characteristics but one thing is for certain, they will be gentle. At first! Swarms are designed to be gentle by nature – as are most small colonies and so you won't actually know what they are going to be like when they get big and ugly. But, by the time the colony does grow, you should by then have the confidence to deal with it. An advantage with a swarm is that you will be handling bees right at the start and tipping them into the prepared hive – or onto a ramp leading into the hive. I repeat, you are handling bees. Your new bees! It gives you confidence. If you decide to go this route, Chapter 5 will show you exactly how to put a swarm of bees in your hive. It's easy!

Some beekeepers say that getting swarms is getting problems and this may be because of bad temper, tendency to swarm, carrying disease and all sorts of other reasons. I believe that these problems are overblown by the critics. Starting with a swarm is one of the most instructive ways of starting beekeeping as far as I'm concerned and later on, an excellent way of increasing your own livestock numbers for free. If you are even luckier, you will have seen the swarm and chased it until it hangs up and then 'captured' it yourself. Brilliant, but first time it's best to grab hold of an experienced beekeeper.

There are two other common methods of obtaining bees:

Packages and 'nucs'

Packages of bees can be purchased and the bees placed in prepared hives but these packages are sold mainly in the American market. In the UK and Europe, small hives of bees (nucleus hives or 'nucs') with perhaps four or five frames are sold containing a queen, bees, brood and stores. The advantage of this method of obtaining bees is that whether purchased from a member of the BKA or from a bee supplier, the queen is (or should be) raised from a stock of known characteristics which she will share with her offspring. In other words, a queen raised from a gentle colony will produce a gentle colony. Usually! It doesn't always work out that way but it mostly does.

When the 'nuc' arrives at your home, simply go to where your prepared hives are and place the nuc hive on top of the empty, main hive with the entrance facing the same way. The simple method involved is described fully in Chapter 5.

Swarms and beehives with bees can be purchased through existing beekeepers and it is very well worthwhile visiting an annual beekeeping association auction or answering ads in the local beekeepers' newsletter available from any association. New bees in nucs or packages can be obtained from any of the bee supply companies or breeding apiaries advertising in the beekeeping press. They will be a fund of knowledge and advice and it is well worth listening to them. Their reputation in a very competitive world depends on your satisfaction and so don't be afraid to believe their advertising or to take their advice and if it is wrong, let others know. It is often a good idea to ask others who may have purchased their bees from the same supplier if they were satisfied with the service.

So what about these bees? What has the beekeeper just put in the hive? Let's now look at these bees and see how they work in a hive.

What are you actually getting?
The bees in the hive

A colony of honey bees consist of three types or caste of bee. The *queen bee* – a fertile reproductive female and usually only one of these is present in the hive; the *drone bee* or male bee which mates with queens and the *worker bees* which do just about everything else.

Honey bees (like other insects) pass through four distinct life stages: *egg*, *larva*, *pupa* and *adult*. This process called complete metamorphosis, means that the form of the bee changes dramatically from the larva to the adult. This process takes 21 days for worker bees. On the first day, the queen bee lays a single egg in each cell of the comb. The egg generally hatches into a larva on the fourth day. The larva is a legless grub that resembles a tiny white sausage. The larva is fed a mixture of pollen and nectar called beebread. On the ninth day the cell is capped with wax and the larva transforms into the pupa. This pupal stage is a physical transition stage between the amorphous larva and the hairy, winged adult. During this stage the pupa doesn't eat and on day 21, the new adult worker bee emerges. Queen bees take only 16 days to complete this process and drone bees take 24 days. These development times, which can vary slightly enable beekeepers to work to definite time scales when working with bees and to plan their operations effectively.

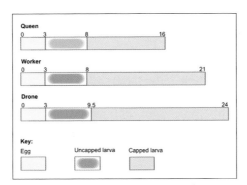

Figure 4: Bee Development Chart. Number of days to emergence.

Queen on emergence: If conditions permit, the queen will make a mating flight five to six days after emergence. She lays eggs 36 hours or more after a successful mating flight, but more usually after three days.

Drone on emergence: Fed by workers until around seven days old. Remain in the hive until approx. 12/13 days old (sexually mature). Thereafter, mating flights during the afternoons.

Worker on emergence: A worker's life span will vary according to the time of year. During the Summer, the average life span is 15–38 days; during the winter it can be 140 days or more.

Queen bee

The *queen* is larger than the worker bees, but not that much larger and is about the same size as the drone bee. Because of this she can often be missed when you are inspecting a large colony, especially as some queens are very shy and shoot down to the bottom of the hive if there is any disturbance. Even experienced beekeepers sometimes have difficulties. When I have difficulty finding her I usually remove my reading glasses and for some reason the larger 'blur' moving about amongst a mass of other moving blurs catches my eye and I mark the spot, whip my specs on again and see her.

The queen honey bee is effectively an egg laying machine and in her prime can lay up to around 2,000 eggs a day during the active season. Without her the colony has no future and if she fails or dies and the bees are unable to raise another queen then the colony will die out. If on ageing, her pheromone secretions lessen, or the hive becomes crowded and her secretions are spread too thinly, then the bees will recognise this and begin preparations to raise another queen to replace her. Swarming may be the result of this and again we will look at swarming and its associated problems a bit later on. The queen has a sting which she uses to sting other queens, a situation which does occur under certain circumstances. More later. Her barb is smaller than a worker's and her sting is better anchored and so if she stings another queen she can withdraw it.

Worker bees

The *worker bees* are incomplete females in that they cannot mate. They do just about everything else though and all of the work in a colony from cleaning, nursing, tending the queen; guarding and foraging for food, propolis and water. And they sting if you mess with them too often. These are the same creatures that pollinate plants, our crops, our wild flowers and trees, our allotments and our fields and gardens.

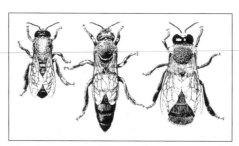

Figure 5: The three castes: The worker, queen and drone bee.

Worker bees' work

When a worker bee emerges from her cell, she will eat pollen which is important for developing her hypopharyngeal and mandibular glands, which will produce brood food and royal jelly to feed brood with. For the first few days she will also help with the housework, especially cleaning empty cells ready for the queen to lay eggs in or for honey or pollen to be stored. After this she will begin a series of other tasks depending on her age. When her food glands have developed, you will see her assuming nursing duties and feeding older larvae with honey and pollen and young larvae – or the Queen – with brood food and royal jelly.

She is able to begin wax making and other duties at around day 10. These other duties include taking nectar from returning foragers and processing it into honey. She will also help with hive ventilation and cooling and some bees may act as undertakers, removing dead bees from the hive. After about day 15–16 she will be able to produce enzymes required for honey production.

About this time, she will spend time guarding the nest and you can see the guard bees on the hive alighting board ready to attack any intruders. Soon she will take orientation flights around the hive and will extend these until she becomes a scout/forager bee. She will fly a total of about 800–1,000 kms foraging unless eaten or killed and will then die. That in a nutshell is the life of a worker bee. All work and no play.

You can see from this that unlike some other insects which have specialist workers or castes used for certain jobs based on physiology such as a soldier ant which looks very different from her nest mates, honey bee castes are age related with bees carrying out the different tasks required by the colony according to how old they are. This is not absolutely fixed however because as research shows, if all the nurse bees in a colony are removed, older forager bees will resume those nursing duties. Now this is quite something because they have to re-develop atrophied food producing glands. Similarly, if all the foragers are removed, a lack of pheromone will make younger bees take up foraging duties earlier than normal. So the colony adapts and adjusts itself for all eventualities.

Another interesting observation is that of course if all the workers became say undertaker bees or ventilation fanners then the system wouldn't work as well so what makes some bees do certain jobs and not others? It is thought that this is where genetic diversity helps. Remember, a queen mates with many drones and so worker bees can be divided into separate sub families within the hive – all with the same mother but different groups with different fathers. These are known as sisters and super sisters. Some groups of workers therefore may be more genetically disposed to certain jobs than other groups. The more drones the queen mates with, the better the diversity.

Drone bee

The *drone bee* is the male of the family. He is the result of an unfertilised egg. In other words, he has no father unlike a worker or a queen and is born as a result of parthenogenesis. Many writers refer to drones as being lazy. This is of course nonsense. He is not designed to 'work' in the same way as a worker bee otherwise he would be one. He is designed to mate and that's what he does. A colony will usually hold about 200 of these at the height of the season. This bee is a very specialised animal which is optimised for mating only. It has no sting, does no work and dies after mating, but a colony without drones for some reason during the main part of the bee year is not happy and works less efficiently. The drone bees are fed and looked after by workers and during the spring and summer they fly to what are known as Drone Congregation Areas (DCA) where they will mate with queen bees. These areas are distant from the apiary. Each queen flying through the area will mate with up to 20 drones and will store their sperm in a special sac or spermatheca for later use. Each drone that mates with the queen dies immediately afterwards as most of his insides are pulled out.

This DCA idea is a neat way of ensuring several things. Firstly, drones from widely dispersed areas will be present thus helping to ensure genetic diversity, and secondly the drones that mate with the queen are the fastest and fittest so helping to ensure that only the fittest mate. Finally, it helps the queen. Flying about outside the hive is dangerous: bad weather, predators, getting lost. All these are common occurrences with worker bees that are often outside. By storing up sufficient sperm for several years, the queen bee need only make this one flight into danger and the drones which don't do any hive work are only around when needed and so don't deplete colony stores of food in times of dearth such as in winter. Drone bees have an amazing array of sensors enabling them to find the DCAs and the queens in flight. They have powerful wing muscles for speed and endurance and they have no sting. How bees determine the location and extent of these DCAs is still a mystery but some known ones have been in the same location for hundreds of years and I have spent much research time testing the light measurements, temperature, wind speed and direction and just about every other parameter whilst watching drones in a DCA try and mate with my artificial queen to try and find out why they are where they are and how they know how to get there.

Some taxonomy and basic bee biology

This bit sounds scientific but is actually easy to grasp and gives you an idea of the truly amazing nature of this evolutionary marvel. Some beekeepers make this aspect of beekeeping their life's work and really become interested in the detail of the biology and taxonomy of bees. Bees are in effect highly modified vegetarian wasps and evolved from their wasp ancestors around 100–150 million years ago when the flowering plants also evolved. These plants and the bees fuelled each others' evolution and just as humans now try and 'evolve' different forms, colours and scents in flowers by breeding them, bees did the same. Bee vision and bee sense of smell modified flowers to suit themselves, just as the flowers modified bee physiology to ensure that they would be serviced by ever more efficient pollinators. More on bee vision below.

Whether this was a partnership of equals or a warring arms race is a much discussed question but the fact is that plants provide honey bees with fuel for flight and energy food to store in exchange for carrying plant pollen from one plant to another; perhaps more efficient and less haphazardly than the wind pollination that many non-flowering plants such as the grasses use. Of course, this evolutionary process took place over many millions of years and may not have finished yet as far as the bees are concerned. Many different forms of modified wasps appeared in this process including solitary bees, bumble bees and ants and all these forms are included in the order 'hymenoptera' which means 'membranous wings'. And yes, ants do have wings but only certain of them, and for a limited and specialised period in their lives, many of you will have seen a mass take off of queen ants looking like smoke from a distance.

Remembering my studies at Cardiff, I recall that taxonomy always seemed to me to be an ever-changing series of complications and if I were you I wouldn't worry too much about it unless it captures you imagination. Below is a chart showing where bees are in the taxonomic order but it is based on morphology or physiology. Increasingly, taxonomists are using biological measures as the basis of hymenopteran taxonomy and so it seems terms are changing. Suffice to say, your honey bees in the UK will probably be a mix of strains of Apis mellifera.

Table 1:   Main division of bees

Phylum	Arthropod	Junction tarsus
Class	Insecta	Divided in head, thorax, abdomen
Classification	Hymenoptera	Hymenoptera
Super-Family	Apoidea	Bees
Family	Apidae	Honey bees and humblebees
Sub-Family	Apinae	Perennially, social colonies
Genus	Apis	Honeybees
Species	mellifera	Western honeybee

Table 2:   Subspecies Apis mellifera – the Western honey bee

Geographic distribution	**Subgenus**
Central mediterranean sea and south-west Europe:	**Ligustica (Italian)**, **carnica (Central Europe)**, macedonia, sicula, **cecropia (Greece)**
Western mediterranean sea and north-west Europe:	**mellifera (North European and UK)**, **iberica (Spain)**, sahariensis, intermissa
Middle East:	meda, adami, cypria, caucasica, armeniaca, anatolica
Africa:	intermissa, major, sahariensis, adansonii, unicolor, capensis, monticola, scutellata, lamarkii, yementica, litorea

In the table above, I have put in bold the bee strains that I have (to my knowledge) used.

Local bees or any old bee?

In the UK, I started beekeeping with a swarm of bees, so who knows what type they were, but they were fairly gentle and fairly yellowish in colour so I guess they were more Italian than anything else – the ligurian bee or *A.m ligustica*. I also used these in New Zealand and then started adding hives of carnica bees. My bees in Spain were *Apis mellifera iberica* – the Iberian honey bee. Very dark to black and very savage, but well suited to dry, hot areas. *A.m. cecropia*, is a gentle Greek bee that you might also come across in the UK. I had a colony of these in Spain and the local farmers called them 'the noble bees' because of their quiet temperament. This gentle temperament differed markedly from that of the local national park authorities with whom I got into a lot of trouble for importing the queen from Greece in the first place and potentially mongrelising the local Iberian strain. They were right of course and even though most Iberian bees would attack me on sight and follow for miles and could well have done with a lot of civilising, they could collect honey by the ton in the driest of situations and that was what they had evolved to do and what they were there for. That for me was also the aim of the game – getting a good honey crop whatever the weather. In drought periods they immediately transferred their attention to aphids on the cork trees and made honeydew – black and delicious and I sold it all. It did teach me a lesson in that with the evidence in front of my eyes, I finally realised that local bees, even if accompanied by horrible traits have evolved into what they are for a reason and are usually the best bees for the local conditions. I didn't make the mistake of trivialising that fact again and looked at the Bee Improvement and Bee Breeders' Association (BIBBA) (see below) website. There's no need to go overboard with this though and as long as the bees you get are nice natured and productive, then you should do well.

Bee talk in the dark

A worker bee unable to dance is rather like a teenager without a mobile phone. Bees use a symbolic dance language that communicates complex information. They use this to tell each other where the best feed areas are and when swarming, scout bees use it to impart information concerning the suitability of nest sites. In the pitch dark of the hive, on a vertical surface, bees perform a dance that describes the position of the sun and directions and distance to a nectar source over a horizontal surface. The direction and duration of waggle runs are closely correlated with the direction and distance of the patch of flowers being advertised by the dancing bee. For example, the distance between hive and recruitment target is encoded in the duration of the waggle runs. The further the target, the longer the waggle phase, with a rate of increase of about 75 milliseconds per 100 metres. The angle of the waggle run from the vertical indicates the angle from the sun that the bee must fly and amazingly, dancing bees that have been in the hive for some time adjust the angles of their dances to accommodate the changing direction of the sun.

Whether this 'dance' behaviour is essential for successful foraging in all areas and all circumstances is debatable but bees that observe the dance can indeed use it to find forage. So armed with this information a forager will fly to the source, and so will many of her mates. They will then return to the hive and dance and another group will fly out until very swiftly your kitchen where you are extracting your honey harvest and have left a window open for fresh air will be full of thousands of bees. I've been in that situation both in a kitchen and in a professional extraction plant where everything you touch stings you because it's got a bee on it. You end up with fingers like marrows. There is a serious point to all this of course. By this method, bees are able to exploit nectar sources rapidly and efficiently and are assured of visiting the most productive sources. It is just another example of just how bees are so successful. There were believed to be two types of dance: the waggle dance which indicated nectar sources at a distance; and the round dance that indicated nectar nearby, but now it seems that the round dance is a waggle dance with just a very short waggle run.

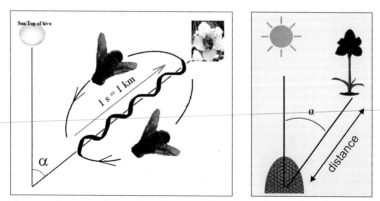

Figure 6: The waggle dance shows direction and distance to forage based on the position of the sun and the duration of the waggle run.

The bee – a very specialised creature

Bee biology/anatomy is a very interesting subject and if studied you will soon realise just what a highly evolved and specialised insect the honey bee has become. The strengthened legs (for carrying pollen loads), the plumose hairs which attract an electrostatic charge in flight thus facilitating pollen adherence, the two sets of wings that are linked when in flight, the multi-faceted eyes that can see ultra violet but not red and that can also visualise the grid of polarised light in the sky that helps in their navigation, the feet that taste, the wax secreting glands under the abdomen, scent glands, the sting – and these are just a very few of the external features of the honey bee. Internally it is just as interesting leading many beekeepers into buying microscopes and books on microscopy.

Bee vision – a different reality

One interesting feature of bees is their compound eyes and what they see. Bee vision shifts the spectrum to the left and so bees see ultra-violet colour as well as blue and green, while humans see a mix of blue, green, and red colours. An example of how this can mean that bees see a very different colour to humans is that many flowers have evolved ultraviolet reflecting pigments, so that whilst a dandelion will seem bright yellow to us, to a bee it is white with an inviting red central zone ready to attract bees to the floral nectary area. Researchers at Queen Mary College in London have actually developed the term 'FReD' – the Floral Reflectance Database – which holds data on what colours flowers appear to be, to bees and other pollinators rather than humans.

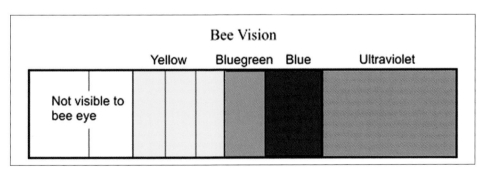

Figure 7: Bee vision. Bees can see ultra violet but not red.
Because of this flower colours reflected in UV light can look very different to a bee.

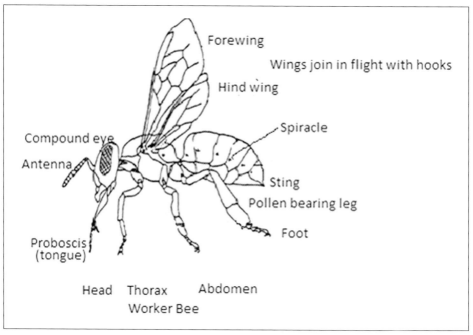

Figure 8: The form of the bee – a typical insect.
The spiracle or breathing tube entrances run along the side of the bee.

The bee in form and function

The picture above gives the main features of the honey bee. Like most insects it has six legs, a head, thorax and abdomen but unlike most other flying insects, the bee has two wings either side. These clip together with hooks while flying. One of the spiracles (mentioned in the text) is shown but the spiracle which the acarine mite heads for is the first one up near the head. The proboscis which is effectively a furry tongue is used for sipping up nectar. The picture is diagrammatic and doesn't show the branched (plumose) hairs on the bee which are one of the features of bees in general to help it pollinate but it does show the thickened hind leg that enables the bee to carry pollen loads.

Pheromones in control

Pheromones control much of a bee's life and the life of the colony depends on them for its smooth functioning. A pheromone is a substance 'released' by a bee which effectively carries a message to other bees that pick it up and tells them to do something or behave in a certain way. For example, when a worker bee stings an enemy it leaves a small amount of alarm pheromone on the victim. Other worker bees picking up this signal will immediately attack the intruder and attempt to sting it in the same area from which the pheromone is emanating or at least to become defensive. In other words, the chemical signal demand has 'obliged' another bee to obey it. There are about 15 known pheromone releasing glands in honey bees and all of them have a role to play in ordering the bee society. Some pheromones have long-term effects such as the queen pheromones outlined below and others have short-term effects such as the alarm pheromone mentioned above.

The queen bee emits two basic pheromones constantly which are essential in maintaining order within the colony and without which, chaos and colony decline would result. Queen mandibular pheromone, secreted by the mandibular gland in the queen's head is often known as 'queen substance'. It affects the social behaviour of the bees and helps inhibit the development of worker bee ovaries. The second important pheromone secreted by the queen is the queen retinue pheromone. It is important for the attraction of worker bees around their queen. This sight of worker bees surrounding the queen can often be seen when you look into a hive. In effect these pheromones are weapons in her quest to maintain her status as the only egg layer in the colony.

Conclusion

This chapter has given a basic insight into the next steps you must take if you wish to continue as well as a look at the basic biology of the bee, a look at how a colony works and your role as a beekeeper. If you decide to come on board, we will now move onto some actual beekeeping.

5
Spring – Your Bees Arrive

A honey bee navigates by using the sun's position, gravitational field direction, and the UV polarisation pattern to determine its direction. The honey bee also memorises any geographical landmarks as aids to better navigation.

The bee year

The bee year for someone just coming into beekeeping effectively starts in May or June. Of course, you can start beekeeping at any time if you buy fully stocked, existing hives of bees, but if you take the advice given here and obtain either a swarm or a nucleus, it will probably be May or June when you receive them.

What you have already done

By now you have hopefully decided that beekeeping is for you. So far you have:

- Read this book and understood the importance of bees to the national life or at least become fascinated with the subject of bees.
- Decided to get involved in actually keeping bees.
- Spoken to beekeepers in the local association and been to several talks and demonstrations as well as seeking advice on the type of hive, and bee you need. Joined the local association.
- Enrolled on and taken a beekeeping course with your local BKA.
- Taken advice and located a suitable apiary site for which you have permission if the land isn't yours and talked to neighbours and family about your plans if you are intending to use your garden or roof.
- Arranged for an experienced beekeeper from the local association to be ready to help you when you set up your hives with bees and to mentor you thereafter.
- Decided (if appropriate) which strain of bee you want.

- Worked out how you are going to install your bees.
- Purchased you kit ready for Day 1 – Bee Day!

You will need:

- Hives on stands (unless bees are coming in existing hives with stands). WBC hives have legs but it is always best to place the feet on concrete or bricks set into the ground to prevent the legs from rotting.
- You will need some hive frames ready assembled with foundation wax. Making up these frames with wax is not complicated but does take some time.
- A good bee suit.
- Good bee gloves.
- Boots.
- Hive tools.
- Smoker.

You shouldn't need anything else at first so if everything above has been sorted, let's move onto Day 1, the day on which you are eagerly awaiting delivery of your bees together with your friendly expert.

Your bees arrive

Now it's time and your bees have pitched up and need a home. It's a good idea now to go through the various scenarios, none of which are in any way difficult. Installing bees is one of the easier moves in the game. I'll assume here that you are not going after a swarm yourself – we'll go into catching swarms (another easy task) later in the book. So here goes:

Scenario 1: Existing hives

Your bees arrive already housed in existing hives. You have already decided where your hives are going to be placed and provided stands for them. Now simply lift the hives from the car or truck (probably with help if they are heavy) and place them on the stands with the entrances facing south east. South east is deemed the best way for entrances to face and there is merit in this because the entrance will face the sun and the bees start work earlier. I have often wondered about this though because I have many times placed rows of hives back to back and haven't noticed any difference in yield. Also I have placed hives on pallets not only back to back but in a horseshoe pattern so hives end up facing in all directions. In many cases, especially in small gardens when you want to place your hives facing a garden wall or fence to make the bees rise over your neighbours' heads, it simply may not be possible to place them facing south east. It doesn't matter. In an urban situation, neighbours are more important than honey yield anyway. Now remember that once the hives are placed, remove the entrance blocks so that the bees can fly. DO NOT forget this.

Scenario 2: A swarm

Your bees arrive in a small box or in a small hive as a newly captured swarm. You have prepared your hive by placing frames of foundation in a brood box. There are two ways of installing your bees in your hive. A show off way and a direct way! For the show off way, invite an audience of those you wish to impress – and ask your friendly expert to stand in the background but within earshot if everything goes pear shaped. It's called the ramp and sheet method.

The ramp and white sheet method

If this sounds something like a magic trick, in many ways it is or at least it looks very impressive. It's almost as though the bees are doing what you command them to do. The impressive way is to place a ramp up to your hive entrance and to be really flash, cover this with a white sheet. Then take the box of bees and tip them out onto the sheet. Hold the box a couple of feet above

A typical swarm box.

The ramp method of hiving bees designed to impress.

the ramp and give it a sharp and sudden shake downwards or give the base of the upturned box a sharp blow. Most of the bees will fall onto the ramp/sheet but many will fly around in all directions. The bees on the sheet will within a very short time start marching up the ramp and into the hive.

Once the first lot of bees have entered the hive you will see many bees

A newly hived swarm beckoning other nest mates by exposing their Nasonov gland (shown white at the rear of the abdomen).

staying around the entrance with their rear ends facing you and fanning furiously. If you look very closely you will see on the abdomen just above the sting area a small white patch.

This shows the location of the nasonov gland which these fanning bees are exposing. With their wings they are blowing what is known as the nasonov pheromone from the gland into the air and telling the other bees that this is the home. 'Come this way and enter'. It is in effect an orientation pheromone. The fanning bees and the white gland areas are very obvious and interesting to watch. Very soon the bees will all have trooped into the hive and the flying bees will also have gone in. Within minutes, the new colony will have sent out scouts to look for forage and other bees will be furiously making wax and drawing out the foundation into cells. Until this is accomplished the queen cannot lay eggs and the foragers cannot store food. Swarm bees are conditioned to make wax comb because until they have done so, they have no home.

The dump method

This doesn't sound so exotic but it is quicker and probably more frequently used by beekeepers. This is how you go about it: Your prepared in advance hive has the usual ten frames of foundation in the brood box. Take the lid off your hive and remove the five frames of foundation in the centre of the box. Take

Dumping a swarm into a hive.

the box holding the swarm; tip it upside down and with a quick jerk, dump all the bees into the space provided by the removed frames. Then give the box a bang to ensure every bee falls out. The bees will end up in a heap on the floor of the hive but will immediately start climbing up the remaining frames.

Now very gently replace the removed frames ensuring that you don't crush any bees. Then close up the hive and leave alone. Easy!

Installing swarms has several advantages over buying hives complete with bees. Small swarm sized colonies are usually very gentle and won't try and give you a pasting every time you glance at them. Large colonies can be ferocious at times. It is best to start with small and gentle. Secondly, swarms are cheap or free. Thirdly, you will be handling and manipulating bees on day one. This gives you confidence. Finally, they are educational. You can on day one see the Nasonov gland business; the frantic building of wax comb; and witness the amazing sight of foragers going about their business in a very short time. There is also the thought that if you invite an audience of non beekeeping mates, you will look a bit of a hero calmly dealing with thousands of stinging, fearsome insects. Just don't screw it up. But on the other hand, better practice a few times first.

You need to keep an eye on swarms for about half an hour after installing them. If the bees very gradually start to cluster somewhere else other than in your hive, it means that the queen has fallen somewhere and has flown up to some perch or may have even stayed on the ground. The rest of the bees will gather around her in a cluster. Simply have another go. It can be more awkward depending on where the bees have clustered but if you can again dislodge them and get them in your hive, they should stay – as long as the queen is with them. Your friendly expert will have seen it before.

Scenario 3: The nucleus

In this scenario you have prepared your empty hives with five frames of foundation in each hive and you are awaiting the arrival of one or two five frame nucleus boxes of the *same frame size as your hives* to arrive. Your five frames will be at the side of the box rather than in the centre so that you can place your nuc bee frames into the centre of your box. When these small nucleus boxes arrive, simply place each nuc on a hive with the entrances facing the same way as the hive entrance. Then immediately remove the entrance cover of the nuc and let the bees fly. Leave this situation until the next day, i.e.

Nucs sitting on hives waiting to be transferred.

a small nucleus hive sitting on top of a full sized hive with the entrance facing the same way. Do not forget to remove the entrance cover.

Sometime the next day, at your convenience – and there is no hurry – it is a simple matter to open up the hive and place the frames with bees from the nuc into the main hive. Shake any bees left in the nuc into the main hive while it is still open and take the nuc away. The bees flying from their new home will re-orientate very quickly and will fly immediately to their new home. When you order your nuc(s) just ensure that if you have National hives, you buy a National sized nuc and so on. Obviously, don't do anything if the weather is atrocious. Just leave it all alone. The bees can easily wait. Then when you have transferred them, close up and that's that. It is not a difficult project and many beekeepers start this way. Just as in using swarms, small nuc bees are very gentle and give you time to gain confidence before they become big, bad and ugly – which of course won't be the case because you will have bought your bees from a reputable dealer who only sells gentle bees with a marked queen!

You have your bees!

Now what? You have bees in your hives and they are already flying well and disappearing off to collect nectar and pollen. It is tempting to seize the moment and take a look in the hive to see what is going on, but don't just yet. The bees in the big hives that you had delivered will be stressed after the move and need time to settle in and calm down. It's no good on Day 1, getting a pasting from upset bees when leaving it for a couple of days would allow them to sort themselves out. The bees are in a similar situation. The swarm bees need time to make wax and comb and you should give them time. I recommend leaving them for a week and as long as they are flying freely and there is no fighting going on or dead bees near to the entrance to the hive, all should be well. It is now simply a matter of taking your bees through to the honey harvest.

First steps for spring

Spring – even late spring is a time when bee populations explode in numbers. You will be amazed at how quickly your new, small colonies will grow. I never cease to be amazed and I always seem to be racing to keep up. One minute they are small and neat and the next, they are bulging at the seams and wanting to swarm. If you have started in the early spring as suggested, this is most likely what they will do and it is this sudden growth and the swarming impulse that we will go into next. So with rapid increase in numbers you need to ensure that your bees have room to expand and this involves certain checks and manipulations. I'll explain build up and swarming and then I'll tell you what you can do about it so that hopefully you won't get caught out and lose half of your stocks that you've only just got. Your bees may or may not swarm. A late nucleus or swarm probably won't in your first year. Remember, you haven't received them until late May/June so it is likely that you won't have to worry about your bees swarming. Yet!

There is another task that you must carry out during the spring time and that is treating your hives against the varroa mite. This small mite officially called Varroa destructor has caused a host of problems for honey bees and beekeepers and is very aptly named. It evolved with the Eastern honey bee (*Apis cerana*) which can deal with it. When it came into contact with the Western honey bee (*Apis mellifera*) there was no defence. It was like influenza hitting Amazonian Indians following colonisation. Apis mellifera simply doesn't know how to handle it and so the mite very soon kills the colony (and so itself) which is not its intention. But we'll come back to varroa later. So, your three main tasks for spring are:

- Ensuring that your bees have enough room in the hive for expansion and food storage.
- Dealing with swarming. (Probably not in Year 1).
- Treating for varroa. A nucleus hive should have been treated prior to you receiving it. Ask the seller if this is so. Don't just assume the bees have been treated.

That's all. Just three (main) tasks, the first two of which are very much linked and in your first year you may not have to deal with swarming and you may be able to leave varroa treatment until August.

Inspecting your bees

These three tasks of course run alongside the usual ongoing tasks that any livestock farmer has of ensuring that you bees are always well housed in dry

Inspecting bees is an important task.

conditions, clear of disease and safe from predators both deliberate such as woodpeckers and accidental such as cows scratching themselves on hives and knocking them over (and wind), as well as flooding and any other harm you can think of. Inspecting your bees is part of this ongoing tasking and in order to ensure that you are doing this quickly and efficiently you need to plan each inspection carefully – before opening up the hive.

What do you want from your inspections?

Books can tell you just so much. For your first inspection, say a week after your bees have been installed, why not ask for a member of the beekeeping association to be with you. Perhaps they have already organised a mentor for you, but if not, ask. I can tell you what to look for during this inspection and will in a moment but someone with you can bring it alive and if you do make an error or drop something then you have someone to assist you on the spot – or commiserate with the fact that you've lost your queen in week one!

Taking a look inside a hive is now becoming one of those contentious issues in beekeeping with many now saying that internal inspections are bad. Well we've been through all that and I repeat here that as long as you are looking after your bees correctly, the internal inspection is not going to stress them too much. I have never used a set inspection routine because often bad weather and other commitments got in the way, but I did inspect my bees about every three weeks except during the swarming season (around April–June) when I tried to make it more often.

When inspecting my hives I always knew what I was looking for before I entered the apiary. I had a set list of objectives which I have listed below and I was also well briefed on any special observations I had to make on individual hives from my records. For example, if I had united a couple of colonies into one I would check their status, or if I had re-queened a hive I would check that the new queen was laying. It's no good forgetting to check something vital and then having to return to the hive and open it up again. If anything is going to get the bees irritated it is this.

So before you actually go and look in the hive, let's look at what is involved in a general hive inspection. What do you need to check? Every check should include the following:

- A queen is present – either see the queen or see eggs.
- There is a good brood pattern.

- There is sufficient room for the queen to lay eggs and for the workers to store honey.
- There are or are not queen cells present – during spring/early summer mainly.
- There are no signs of disease.
- There are sufficient stores of honey for the bees especially during those times of the year when there are few sources of nectar available.

That is the basic check and all of these points should be checked out every time you look into the hive. You can then learn to make decisions based on what you have seen. Note that all the points except the last point are concerned with what is going on in the brood box and so we will now look at this in more detail and at the end of the chapter is a full hive inspection check list.

Checking the brood box

The brood box is where it all happens in a bee hive. When you look at the brood nest, you are checking up on the queen and this can be done whether you see her or not. This larger bottom box is where the home is, where the queen lays her eggs and where the new bees are born and nurtured. Inspecting this area provides a beekeeper with a visual historical record of what is going on now and what has gone on in the past. A frame of brood with its arcs of sealed brood, larvae, young larvae and eggs surrounded by honey and pollen cells is a history of the colony. How is this so? It's a matter of simple arithmetic and if you take a look at Figure 4 (page 54) you will see what I mean. This diagram shows clearly the timings (in days) involved in the development of eggs, larvae and sealed larval/pupal stage. Remember this as you go through your inspection schedule, but for now, let's go through the list.

When you take a frame out from the brood nest, you are looking at a very efficiently laid out, living part of the home. You hopefully will see a central mass of good sealed cells – sometimes with the odd empty cell amongst them. An arc of unsealed cells with pearly white, coiled larvae should surround the central area. Further out will be an area of younger larvae, still in their beds of royal jelly and further out still should be an arc of cells containing eggs. This area is often surrounded by arcs of honey and pollen filled cells. It isn't always that perfect and there are very many variations. For example you may pull out a frame that is completely filled with sealed larvae or one with just about all eggs.

Capped brood surrounded by unsealed brood of decreasing age is a normal pattern.

Is the queen present? In a small nuc or swarm with perhaps four or five frames this is not too difficult. You will notice much more easily if she has a painted dot on her thorax Paint colours vary according to the year with the following code usually implemented:

Year	Colour
0 or 5	Blue
1 or 6	White
2 or 7	Yellow
3 or 8	Red
4 or 9	Green

So a queen born in 2012 would have a yellow dot on its back

This marking system has many uses, especially for research or queen breeder use but personally, I always painted my queens white whatever the year because it was easier to see. Marking queens can be tricky but marking kits do exist and are available from bee supply stores but if your queen is not marked, ask a member of your BKA to show you how.

If you see the queen, also look for eggs. These are small, white little stick like things standing upright right at the bottom of the cell. If you've seen

The queen can live for up to about four years,
whereas her worker sisters will die after a few weeks.

Bee eggs. Upright and one to the base of each cell.

the queen and also seen eggs, then you know that she is not only there but laying.

In larger colonies, the queen may not be found so easily. There are thousands of bees milling about, some of them having a go at you and finding the queen hidden in that lot can be daunting. She will likely be on a frame with brood on it, so look at these first and search the entire brood box by lifting each frame slowly gently and looking carefully at each side of the frame. Watch out for a bee moving usually more slowly than the others, often moving towards the top of the frame so as to go over to the other side. To my knowledge there is no sure way of seeing her but as I mentioned earlier, a white dot helps so ask your supplier to supply a painted queen if you buy a nucleus hive. You may have seen pictures of the queen surrounded by her 'court' of attendants. I've rarely seen this except in photographs but am told that it is a fairly common sight. Usually I see the queen marching about on her own, often trying to get over to the other side of he frame I'm holding, so don't expect the ideal.

Once you have seen her, and seen eggs so you know she is laying, place the frame with the queen back into the hive carefully and slowly. Don't drop it. You know now that the colony has a laying queen which is the very foundation of colony life.

Can't find the queen?

But what if you can't find her? Some queens are very difficult to find especially in large colonies where every bee seems to look the same and you keep seeing drones as queens. Some queens are very nervy things and dive to the bottom of the hive if disturbed. So what now? I have heard of many ways of finding her, some more or less valid but many, rather useless. One method involved sieving the bees through a queen excluder until the queen was found. This is the sort of thing that would really stress up a colony and really is unnecessary.

Firstly, look for eggs. If there are eggs, then the queen was there within the last three days and is probably still there and so it is not vital that you find her. Personally I still like to find her especially if I haven't inspected that hive for a long time (as can happen), just to make sure that she has a white dot on her thorax. If she hasn't, then there is another queen there and I want to know that so that I can record the matter in my note book and get a better picture of my hive in my mind. If I simply can't find her and I see eggs and a good brood pattern then fine. It is not worth taking the hive apart and possibly damaging or losing her and upsetting the bees.

If there are no eggs, then look for very small larvae still sitting in a bed of royal jelly. If they are there, then (remember the chart at Figure 4, page 54) you had a queen within the last eight days but may have lost her three days ago. If you see good sealed brood only this shows what was happening nine to 21 days ago and it means that you may have lost your queen or your bees have swarmed and the old queen has gone leaving a young virgin behind who has yet to mate, or if mated has yet to start laying. There could be several reasons why you can't find her or why there are no eggs and it may then be a good idea to ask a more experienced beekeeper for help on this because you need to make the right decision as to what to do about it and there could be several different possibilities.

Is there a good brood pattern?

You have already taken a look at the frames of brood so was there? The typical brood pattern is described above but as I mentioned there are variations. One good pointer is to look at a frame of mainly sealed brood. Are there lots of empty cells making the frame look like the lid of a pepper shaker? This is a sign of inbreeding and if repeated over all of your brood frames then there is likely to be a problem with your queen. Again, seek advice from your mentor or beekeeping association.

Is there sufficient room for the queen to lay eggs and for the workers to store honey?

This is an easier one. Unless the colony is heading towards winter, then the queen needs room to lay eggs. If you see that the brood box is bursting with bees and that every frame is full of wither brood or honey, then the queen needs more room. If you are using a queen excluder, remove it, place another brood box on top of the existing one(s) and replace the excluder on top. You shouldn't get to this stage because if your boxes become too full like this in the spring, your bees will probably swarm and would have done so by this time. On your first inspection, this would be an unlikely scenario but remember what I said about springtime. The bees really expand at an incredible rate and you will need to add brood and honey boxes very soon. For a brood box, add another after about seven or eight frames have been filled. Honey is also stored at a fast rate in the spring and you may need to add honey supers after about seven frames have been filled. Some may ask, well why not add five honey boxes immediately and just leave the bees to it. The reason is that in this case, the bees, which tend to work upwards will do just that. They will fill the centre two or three frames of the lower box and move upwards and do the same in each box. Then when you harvest the honey half of the boxes will effectively be empty. I have seen and done this and found out the long way that it's best to put each box on one at a time when the top box is three quarters full. You can place the new honey box over or under the top box.

Are there queen cells present – mainly during spring/ early summer?

This is unlikely to be the case in a small colony with a young queen. However if you have purchased secondhand hives complete with bees, then this might easily be the case and it is an important thing to look for. If there are queen cells present, then the bees have either swarmed or are about to. Having looked at hundreds of colonies with queen cells, I found that mostly, if they had cells,

A well-developed queen cell. Note the well defined sculpting.

swarming had taken place. In other words, your queen has gone and the new queen has not yet emerged or if she has, the bees are waiting for her to successfully mate before destroying the rest of the cells. A sure sign that something has happened is if you look for your painted queen and find an unpainted one marching around.

You find queen cells. Now what? Firstly, it's no good destroying them all until you know whether you have another queen or not. If you do see another queen in the hive, you must then look for eggs. If there are eggs, then you can safely destroy the remaining queen cells although I would be surprised if this hadn't already been done by the bees. Look everywhere for them anyway just to make sure and destroy them with your hive tool.

If there is a queen but no eggs then it is likely that the queen is a virgin and has not yet mated or that she has mated but not yet started to lay. In this case, close up the hive and look again in a week's time. By this time the queen should have mated and started laying. Look for eggs. If you see eggs, all is good.

In your first year, I doubt you will have this problem. In all likelihood you will see a perfect brood pattern with eggs, larvae and stores and a painted queen marching around as though she owned the place and no queen cells. Hopefully!

Are there any signs of disease?

This is sometimes not an easy thing to ask a new beekeeper. Some diseases such as American Foul Brood (AFB) which as you will see in Chapter 10 are absolutely deadly, are very difficult to see until it has progressed. Even very experienced, professional beekeepers miss it, even when deliberately looking for it. European Foul Brood and other diseases can also be very difficult to spot early on. So what can you do? The main way to recognise a problem is to know what a normal brood frame looks like. The sealed cells should be slightly convex and all the same, even colour. The bigger larvae should be pearly white and evenly coiled with no discolouring or strange appearance. The small larvae should be floating in their bed of almost translucent royal jelly and the eggs should be standing proud at the base of the cells. There should be an even graduation between each type and there should be no sign of any moths in the brood box. And it should smell clean and waxy. You will know the smell from apiary visits. If anything is different to all that and/or if the brood box smells bad, then you must seek advice straightaway. In my early days I left things in one hive in the hope that the bees would sort it out and it would just go away. It didn't and I lost the hive and became a wax moth keeper (see Chapter 10).

Are there sufficient stores of honey?

Are there sufficient stores of honey for the bees especially during those times of the year when there are few sources of nectar available?

This is the last of the standard checks but no less important. Bees survive times of dearth for example winter, or high summer in many countries by storing honey. They have evolved this method of survival over millions of years and have evolved methods of protecting it from both bacteria and humans. If you take their honey away after the spring flow or in the autumn and there are no more honey sources immediately available and you don't feed them, they will starve. It's as simple as that. We go into more detail on how much your bees will need over a normal UK winter (if there are any now) in Chapter 9, but suffice to say that if you see your bees have no honey, feed them. It's not difficult. Simply mix 1 kg of sugar to 1 litre of water, mix well and pour into a frame feeder and place it in the brood box with plenty of bits of wood floating on the sugar so that the bees don't drown. Feeders are fully described in

A good frame of honey nearly 100% capped.

Chapter 9. Repeat as required. I prefer to ensure that the bees live on honey not sugar syrup and so if you know your flows or can obtain advice from a local beekeeper you should be able to plan your bee year well and make sure that the only time you have to feed your bees is in a genuine period of drought or an extended winter, when there is no other choice. In your first season in the spring you should not need to feed and if your bees run out of stores then it is because they are in an area with no sources of nectar. Having taken advice from your local association that is unlikely to occur.

What about pollen?

Now for a 'feed' point that even some more experienced beekeepers fail to take on board. Bees need pollen sources. They need pollen as brood food. It provides them with protein. If there is no early pollen, colonies will not build up well in the spring. Make sure of your local pollen plants. Gorse, alder, blackthorn, willow, rock roses are all good early pollen sources for bees. Your local association will know all about this. I know beekeepers who haven't seemed to understand this point and when their colonies haven't prospered, they have blamed other causes and made wrong decisions in treating the problems. It is unlikely that you will need to feed pollen patties but again take local advice and if necessary buy some pollen supplement feed from a bee supplier.

That is the standard checklist for a normal hive and if you wish to avoid swarming then inspections should be carried out from early spring to summer about once every two weeks. By basing your inspection times on queen development times (see Figure 4 chart again) and destroying queen cells during these inspections, many hope to prevent swarming. As far as I'm concerned there are better and less intrusive ways of doing this as we will see. I wouldn't bother with his method. It works against the bees, not with them.

The inspection

Armed with the information above you can now carry out your first hive inspection.

Choose a pleasant, warm, calm day and follow the procedures outlined below:

- Get your smoker going well and try and produce some thick, cool, smoke. Some beekeepers may find this difficult and if it's windy it can be. The way I do this is to firstly roll up some corrugated cardboard into a tight roll. I have these rolls prepared and ready. I light some newspaper and place this into the bottom of the smoker and I gradually lower the cardboard onto the flames, gently puffing at the same time. The cardboard catches and I close the lid still puffing. Vey soon, the smoker is sending out thick cool smoke.
- Give a couple of puffs of smoke into the hive entrance.
- Stand at the side of the hive. I do this because it is out of the flight line and gives me a better view of the frames as I lift them out. I have always place my frames front to back in a hive.
- Lift off the hive lid slowly and gently smoke the bees along the frame tops. Place the lid upturned on the floor out of your way. If there is a crown board then remove that and place it in the lid. You may have to lever the lid and crown board off with your hive tool and this can be done smoothly and gently. As you lift the lid or crown board, gently smoke the bees. A couple of puffs will do.
- If the hive has honey supers on it, remove them and place them on the upturned hive roof (unless it is a WBC gables roof in which case you won't

be able to). I used to take along an empty box to use in this case. Then puff some smoke over the bars of the brood chamber and remove the excluder if you are using one.

- Then remove one of the frames next to the side wall of the box. Lever it out with a hive tool if necessary. Look at it carefully to make sure that the queen isn't on it and then stand against the side of the hive. This gives you room to move the other frames around and lift them out without rolling bees against the adjacent frame. Then gently separate the frames and lift out the centre frame. Look at one side and then turn it around and look at the other side. Before I do this I look over the top of the frame and then turn it around end to end. Look for all the features described above: queen, eggs, open larvae, sealed larvae, stores. If all is well, place it back and remove one of the frames to the side. I understand that the BKA teaches beekeepers to inspect from one side of the box after removing the first frame and this is a logical and perfectly good way of doing it. I got into the habit of starting in the middle and working outwards a long time ago because it usually (not always) enabled me to find the queen – and I usually like to see the queen. Either way, try and be consistent so as to make your record keeping easier.

When you have finished with the brood box, replace the frame at the side of the hive and place the excluder back on top. Then go through the honey supers just to check if there is any and if more room is needed. Then put everything back together and that is your hive inspection. None of that was difficult and next time with a bit more confidence you can spend a bit longer and watch bees dancing or bees emerging from cells and you will gradually begin to understand more about the workings of the hive and you will slowly and surely begin to 'read' the hive.

Keep a written record

Make sure you keep a record of each hive that you have inspected. In this way, you can build up a picture of your colony. Don't think you will remember without writing a record because you won't. I always check the following:

- Queen seen (paint?).
- Presence of eggs.

- Presence of unsealed brood.
- Presence of sealed brood.
- Good brood pattern.
- Sign of disease.
- Sufficient stores.
- Signs of congestion/over population.
- Presence of queen cells.

I also record aggressiveness/calmness as over the years it has given me a good idea of what to expect in this respect. The BBKA produces a hive inspection template which is a good way of recording what is going on in your hive.

Interpreting honey flows

It is after an inspection like this that an experienced beekeeper who knows his bees and the local floral nectar flows will think: 'I've got a couple of boxes on each hive full of early honey here. As it's mostly apple honey and I know that soon there will be a flow from the local beans or even oil seed rape, I could harvest this lot now and wait for the next flow and then harvest that in the late summer. The bees will easily last until the next flow as long as I leave them something.' But unless you have this knowledge, you'd best leave the honey well alone for the moment. Use this first spring and summer to learn about the different flows from the local resources. Make notes and watch everyday which direction your bees are flying in. True, every year isn't the same, especially in recent years but time marches on and you'll be surprised at just how quickly you build up your local knowledge even with the vagaries of nature involved.

Conclusion and summary

This chapter has started you actually beekeeping, installing your hives and carrying out hive inspections. I hope that it has described just what to expect during spring and that it has reassured you that beekeeping is not that difficult – just extremely interesting.

INTERNAL INSPECTION WORK SHEET

- Puff some cool thick smoke into the entrance of the hive.
- Stand to one side of the hive and lift the lid. Puff some smoke over the top bars of the box. (Remove the crown board if there is one). Place the lid upside down on the grass.
- Lift off all the supers and place on the lid in the order you find them.
- Remove the queen excluder from the top of the brood box(es).
- Puff some smoke over the frames. Gently lift out the outside frame from one side of the box. Ensure that the queen is not on it and place it to the side of the hive.
- Now lift each frame out gently and look for:
  - A queen is present – either see the queen or see eggs.
  - There is a good brood pattern.
  - There is sufficient room for the queen to lay eggs and for the workers to store honey.
  - There are or are not queen cells present – during spring/early summer mainly.
  - There are no signs of disease.
  - There are sufficient stores of honey for the bees especially during those times of the year when there are few sources of nectar available.
- If you find that the brood box is congested and that there is nowhere for the queen to lay, then add another full or half box.
- Note every observation down and if there are deficiencies act on them or plan to act on them if not urgent.
- Replace all the frames and replace the queen excluder.
- Check each super for stores and if the top super is three quarters full and there is a honey flow on, place another super on top.
- Replace the supers on the hive having noted down the amount of stores. Super if required.
- Close up, strap up if necessary and move to the next hive.

Now look at your notes and take action if required on all the points noting deficiencies. This book will tell you how.

6
Spring to Summer

*Honey bees have an unusual genetic sex determination system
known as haplodiploidy. Worker bees are produced from fertilised
eggs and have a full (double) set of chromosomes. The males, or
drones, develop from unfertilised eggs and are thus haploid with only
a single set of chromosomes.*

In an ideal world, once you have installed your bees and know how to inspect them, you should now be able to leave them all alone until you obtain your harvest of honey. But of course livestock farming just isn't like that. You have to manage your charges and help them work your way by working their way and this is what this chapter is all about. Hive management. In your first year, some of these problems may not arise. For example, swarming should not occur if you obtained a small nucleus or swarm in May or June. (Although I have had swarms, swarm within six weeks of capturing them because of their phenomenal rate of build up so it is best to be prepared and take all precautions.) Hopefully however, you can wait until the spring of your second year, so keep this chapter handy. If you purchased a nucleus, then it should have had its spring treatment for varroa and so you can leave treating your bees until the autumn.

Varroa destructor

Varroa destructor is a fact of life and something that has to be dealt with. It is a mite that evolved with Apis cerana, the Eastern honey bee which can tolerate it and it jumped ship to *Apis mellifera,* the Western honey bee that can't. In fact if you leave varroa infestations alone, your hive will die out. That obviously isn't the intention of the mite but it is reality. There is more on varroa's effects on the hive in Chapter 10 because they are far ranging but for now it is sufficient to know that you should treat it both in the spring and in the autumn. There are several ways to treat this beast. You can buy proprietary treatments at any bee supply shop and these consist of thin strips of a special plastic impregnated with a miticide. You simply insert these between the frames according to the instructions on the pack and leave for the required number of weeks. They are effective and easy to use but there is a problem and that is that the mites are slowly becoming resistant to the chemical employed. In some areas of the world, this is a very serious problem. For this reason, many associations now advise that you use this type of chemical treatment in the spring and use organic treatments in the autumn. This would have the effect of slowing down the build up of resistance. An alternative is to use organic treatments all the time. I did. I was an organic honey producer and because off-the-shelf organic treatments were so expensive, I used oil of thyme in an olive oil base. However there are organic treatments available for purchase which use more measured doses set in gels providing a more stable and measured release effect.

There are other treatments available as well such as using oxalic acid, sugar or dust and so on and these will be covered later for interest, but my advice for your first year is to seek the advice of your beekeeping association. This also has another benefit. If everybody treats their varroa infestations at the same time it is hugely beneficial for the whole area. It just takes an apiary of bees that have been left untreated to cause a re-invasion of your treated hives. Purchasing organic treatments for just two hives would not be expensive.

Swarming

This is quite a subject and one that is fundamental to bees and beekeepers. Swarming is reproduction at a colony level by non-sexual means and there are many theories as to its cause. It is probable though that when the hive becomes overcrowded, the amount of queen pheromone available per bee becomes less especially in areas such as the bottom bars in the hive away from the centre. In response to this, worker bees start to build what are known as queen cups, small acorn shaped cells jutting out from the main comb. Construction of these doesn't necessarily mean that they will build these into queen cells but it can be an indication. If swarming preparations continue and the bees perceive a need to swarm, then the queen will be manoeuvred into laying worker eggs into the acorn shaped cups and queen cells will be built from them. These are beautiful objects looking like well sculptured peanuts in their shells hanging down from the comb or from the bottom bars. There could be many of them.

Once this has happened, then it is likely that the old queen will fly from the nest together with around half of the bees and set off to find a new home. Every bee in the swarm will have sufficient honey in their stomach to start building wax comb for the new home and if the scout bees find a suitable cavity, they will head for it and start anew.

Meanwhile back at the ranch, a new virgin queen will emerge from one of the queen cells. (She may have already done so before the old queen goes.) She will probably kill the other queens in their cells by stinging through the cell walls. Sometimes however the bees will protect queen cells from this and another virgin may emerge. Unless the bees protect this second virgin there will be a fight and one will win. The bees often protect another virgin just in case the first flies off to mate and doesn't return. The winner or the remaining virgin will after a couple of days fly off to a DCA and mate and will hopefully return to the hive and spend her life laying eggs as what many commercial beekeepers term 'a production unit'! She won't ever mate again but may fly out with a swarm when her turn comes. Thus where there was one colony, there are now two. Reproduction! And because it is reproduction it is very difficult to stop it happening.

Why stop swarming?

Why should you want to stop it? Because half of your livestock disappears and very often legally becomes some other beekeeper's livestock. This can mean that you will get no surplus honey that year. It can also upset your neighbours if you are an urban beekeeper. I mentioned earlier that one way of trying to stop this happening is to kill all the queen cells yourself. Usually though one or more well hidden cells will be missed and the colony will swarm anyway so this isn't really a foolproof method.

Clipping the queen's wing (don't do it)

Another way is to use a queen that has had one of her wings clipped. Personally, I cannot think of a worse way of going about beekeeping. The theory is that she can't fly out and so neither will the swarm, except that what often happens is that the swarm goes, she also tries to fly out and is lost. I read one account supporting this practice of wing clipping that said that when she falls to the ground near the hive, she will walk back to the hive! I laughed out loud. She is more likely to get lost or eaten or both by a myriad of predators. I have been castigated in the bee press by the elders of the beekeeping establishment for saying that this method of swarm prevention is no good and I probably will be in the future. But still, my advice is don't do it. As I said earlier on in the book, if you look after your bees properly, they will not be stressed and many of the problems that are facing bees now can I am sure be avoided. Clipping wings does not constitute looking after your bees.

So what can you do instead of mutilating queens? You can try and prevent the conditions that lead up to swarming in the first place. If you have purchased a small nucleus or a swarm then it may not happen anyway in the first year – but it can. I've seen it. Let's now look at the signs in the hive that may lead to swarming and then take a look at preventing this.

Signs of swarming

From March onwards (and getting earlier) you should look for preparations and signs of swarming behaviour. Some indications are as follows:

- An overcrowded brood chamber with no room left for the queen to lay eggs.
- Lack of space in supers for honey storage.
- Laying of eggs in those acorn shaped queen cups mentioned earlier.
- Building of queen cells.

If you are really observant, then the following less easily visible signs can be apparent:

- Weight loss of the queen in preparation for flight. (up to $\frac{1}{3}$ to $\frac{1}{2}$ of body weight).
- The field bees do less work and may congregate at the hive entrance and/or on the lower frames.
- More drones are reared.
- The queen lays fewer eggs.
- Prior to leaving the hive, the workers engorge themselves on honey and nearly cease normal flight activity. This is on the day.

Then suddenly there is a mass of bees charging out of the hive and swirling around in the air above the hive. It will soon start to move off. If you happen to be there you can follow it and capture it. More on this later.

Methods and manipulations to help prevent swarming

According to one famous American beekeeper the most effective way to prevent swarming is to dynamite the hive. He does go on to say however that if you did that, you probably wouldn't get much honey out of them that year! Yet in a way he had a serious point. It is difficult to stop unless you resort to total destruction and with some bees, whatever you do they will swarm regardless, so if it happens and you've done your best, accept it (and try and retrieve the swarm). Here are some methods to help you avoid this situation.

Use young queens

The use of a young queen is a good way of avoiding swarming so if you re-queen each year or every two years your bees will be less likely to swarm. The difference in swarming tendency is marked and the figures are quite dramatic. This is another reason why you probably won't be troubled by swarming if you start your first year with a nucleus hive and its new queen. If you start with a swarm or an existing hive you may have less luck.

Reversing hive bodies

This is best done if you see queen cups being built in numbers. Bees tend to work upwards and often after winter they are in an upper box and tend to get jammed in there not realising that they have loads of space below. Simply swap the two boxes and repeat this as required. Even if you have put on honey supers, this can be done. It's easy but should be carried out with other methods.

Equalising colonies

There are two ways that you can do this. Firstly, you can move frames between colonies, or you can exchange the positions of the colonies and this manipulation assumes that you have a colony that has built up slowly and is weak in bees and one that is bursting at the seams. Anyway, here is what to do.

Exchanging frames

If you have a very full colony wanting to swarm and a weak colony that is building up slowly, transfer frames of brood from the strong colony to the weak one and exchange them with frames from the weak colony. Basically you are building up the weak colony to give you a honey surplus and relieving the pressure on the strong colony.

Dangers!

- If your weak colony is weak because it has a disease, then you will be transferring the disease to the strong colony. Not good. You may need help with this from a more experienced beekeeper.
- You might accidentally transfer the queen, so make sure that each frame of brood that you transfer has no bees on it. The weak colony taking the full frames of brood is relying on that brood to hatch and boost it up – not any accompanying bees that might in any case fight in the new hive. If the weak hive is very weak, there may be insufficient bees to look after the new frames of brood and they will get what is known as chilled brood and die. Keep an eye on this. Make sure that each new frame full of brood is soon covered in nurse bees tending to the larvae or keeping sealed brood frames warm.

Swapping hive positions

Again, this depends on you having a weak hive and strong hive. Simply swap the positions of the hive. All the foragers from the populous hive will return to the weak hive thinking it's theirs and will boost numbers. Foragers returning with honey or pollen will generally be accepted in a strange hive so there shouldn't be any problems there. Again, you must make sure that your weak hive has no disease.

The manipulations above were easy to accomplish and now there is another one that I will now describe. Don't be afraid just because you now have to delve into the hives and shift things around. And if you are doing this for the first time, do try and get an experienced beekeeper to help you.

Again, if you have purchased a nucleus hive or a swarm, you probably won't have any problems during your first year but if you purchased hives with bees then you need to know about this method which is a basic beekeeping manipulation anyway and will stand you in good stead for a whole range of other activities in the future.

The artificial swarm

Now what you are doing here is pre-empting the bees and getting them to swarm under your control. You are working with the bees in the direction that the bees want to go. So if you find queen cells or the small cups with a single egg or larva in each of them in a very crowded hive, just carry out the artificial swarm. You will need another brood box preferably with frames of drawn comb but a mix of foundation and comb will do. It goes like this:

- Prepare a new site for the hive near the old site. About a yard or two away. This is site B.
- Move the hive in question to this site B.
- Place the spare brood box on a floor on site A. Have a lid ready.
- Now go to the full hive on site B and look for the queen on a brood frame. This is the only awkward bit. You do need to find her.
- Place this frame with the queen in the new hive on site A.
- Fill up the new box with frames of comb/foundation. I generally place another frame of brood and a frame of stores into the new hive and then I hold one more brood frame from the hive above the new hive and shake all the nurse bees off it into the new hive.
- Place any supers from the old hive plus a queen excluder onto the new hive.
- Place the old hive anywhere in the apiary. The foragers will fly to the new hive on site A.

Then:

- Cut out all the queen cells in the old hive.
- One week later again cut out any new queen cells in the old hive except one or two. One of these will become a new queen and you will have another hive. If you don't want another hive then you can wait a couple of months and unite them. More of that later but it's another easy task.

So what you have done is formed two from one which is just what the bees would have done if they had swarmed but you did it under your control and you didn't lose any bees.

I have described six very useful methods of limiting swarming in your apiary and the last method, the artificial swarm is usually foolproof. Each method works with the bees rather than against them and are usually successful. There are other more advanced methods involving the artificial swarm and you can find all of these in the *Beekeepers Field Guide* – a beekeeper's operations manual from the same publisher but the aim here though is to equip you with the knowledge to get you through your first year May/June to May/June.

Uniting colonies

I mentioned that if you didn't want an extra hive of bees, then you could unite the two colonies. This is easy to do but remember that bees from one colony will recognise that the bees from the other colony are not their own and so they will all fight. You must stop this happening and the way to do this is to introduce them to each other over a period of time. In this way it is thought that their odours mingle and in the end they merge peacefully.

The method I recommend goes as follows:

- Locate the least populous hive and remove its lid.
- Place some newspaper over the top box and to keep it in place use drawing pins to fasten it down. The whole of the top of the box must be covered.
- Make a few slits in the newspaper (not too big) with your hive tool.
- Then remove the floor from the other hive and place it bodily onto the hive covered in newspaper.

The bees will chew away the newspaper overnight and by the time they have got through it, they should have mingled. This generally works without any problems. You may find a few dead bees around but not many at all.

Some say that it doesn't matter which hive you place on top, but once I placed a hive with few bees in it on top of a populous hive and in the morning found the bees in it still up there near the lid. I'm not sure why that is but since then I have always placed the larger hive on top. I'm not sure if it makes any difference.

You loose a swarm

What happens though if despite all your work, the colony swarms? Most likely you won't be around to see this but you just might or a neighbour will ring you because you are the nearest beekeeper around that they know and say that a swarm is hanging up from a branch of their apple tree. It's usually more awkward than that but let's be positive. In this case you don't really know if it's yours or not until you get there and see the queen with her painted dot but anyway, if you can get it, do so. It's free. Swarm bees are full of honey ready to make their new home and so are disposed to be gentle unless severely riled so it's quite a safe business.

Make a swarm catching kit

I have a small swarm catching kit always available which consists of a nucleus box with a mesh floor and no entrance. This is easy to make and those of you who purchased a nucleus of bees will have one ready. Even if it doesn't have a mesh floor it will work, but make sure the entrance is blocked. I also have four or five frames of foundation available. This together with bee suit, a hive tool and a smoker makes up the kit. That is all you need in an emergency. A feeder, frame of comb and other non-essential but helpful items can come later.

Catching the swarm

Go along to the swarm, hold the nuc box underneath it and gradually move it up and over the swarm if it fits. Then give the branch a hard knock downwards

Catching a swarm in a bush.

and all the bees will fall in a heap to the floor of the nuc box. Now place the box on the floor and place the lid on it ensuring that there is a gap so that the bees can go in and out. If the box hasn't got a lid, place it upside down on the ground – the bees will cling to the 'ceiling' of the box.

Now watch for ten minutes. If the bees gradually leave the box – and it is a very gradual movement, and hang up on the branch again, then it means that the queen didn't fall in the box and she has flown back up to the branch. The bees are joining her. If this happens, just repeat the procedure. Then watch again. It is most likely that all is well and the bees have now occupied the box as their new (temporary) home.

Be careful

Unfortunately swarms can hang up in very awkward places and I'm afraid that no specific advice can be given on this except to use your initiative. Swarms on posts can often be smoked upwards into a hive box with frames of comb. Just hold the box above the bees and gently smoke them. They will move away and upwards towards your nice wax frames. Also, be careful. I once went up a ladder to get a swarm fairly high up in a tree. I held the box under the bees, banged the branch with my free hand and the bees fell into the box just as the ladder fell away. I was left hanging onto the branch with one hand and holding a box of bees in the other and you will be surprised how heavy a swarm is. Needless to say I fell into the shallow stream below and was covered from head to foot in bees that were by this stage getting a bit fed up with me. I went back later but they had gone. A friend of mine tried to gather a swarm from an electric fence with hilarious results because he tried three times before he realised that the pain was from the electric wires and not the bees. That was interesting to watch. Another colleague was coaxing a swarm on a fence post up into a box when two bulls having a savage fight crashed through the electrified fence knocking him flying and scattering bees and hives in all directions. Interesting things, swarms and most beekeepers will have 'swarm stories'.

Once the bees have settled in the box and are coming and going happily, just leave them there until nightfall. After dark, close up the nuc, put it in the car and take it home. Either place it in a position where you will make up a hive the next day, or place it on an empty hive with the entrance facing the same way as that of the hive.

The next day, remove the frames from the nuc and place them in the hive. If you have just made up a new hive, place it on the very same spot as the nuc.

Swarms can end up in awkward places.

If you know which hive the swarm came from, don't put it back in that same hive because the bees will be in the same situation that caused them to swarm in the first place and they will do so again. Work with bees, not against them.

The hanging swarm. What is going on?

When the bees hang up in the swarm after having left the hive, what exactly is going on? The swarm is waiting for scout bees to find a suitable home for them. The scouts will identify several of these and will return to the hive and dance on the face of the swarm giving information as to suitability, direction and distance. Dance followers will then go and checkout these various sites and will come back and dance for that site. If the dances describe too many sites or the proceedings become prolonged, scout bees that have been to previous sites will head butt the dancers and cause them to stop dancing. The most vigorous and numerous dancers for a particular site will normally win out and once a site attracts a 'quorum' number of scouts, the bees detect it, and begin to change their signals on the swarm. They then produce a piping signal by vibrating their wing muscles while pressing down on another bee. This signal leads the swarm bees, most of which simply hang quietly in the swarm during the decision-making process, to warm up in preparation for takeoff. At this point, the message of the stop signal changes, and can be thought of as, 'Stop dancing, it is time to get ready for the swarm to fly, and guided by the scouts the swarm will take off.' The scout bees will move through this swirling mass of bees in the direction of the new site and will guide the swarm to their new home. Although swarms appear to be just a mass of bees swirling in all directions, they actually move quite fast as you will find out if you try and follow them.

Catching swarms is a good way of increasing your stock of bees and does provide a service to the community, especially an urban community. One way you can attract swarms is to use bait hives

Bait hives

These are easy to set up. Simply use an old brood chamber, floor and lid and place it about a metre off the ground somewhere. Place a few old brood combs in the box as well as some frames of foundation and leave it be. Scout bees looking for a new home will be attracted to it and investigate it and if they find it suitable, a swarm will occupy it. This could even be a way to start beekeeping in the first place having borrowed the box and comb. Swarm lures can also be used for this and they are based on the nasonov pheromone which will attract the scouts and the swarm, but research shows that old comb is just as effective – and it is cheaper.

One possible issue with this strategy is that you must locate the bait hive on the spot you want to leave it after the swarm has entered it because if you don't notice that a swarm has occupied it for a few days, it is then going to be a problem moving it to where you want it. Unless that is you want to move it more than two miles. Moving bees is a subject in itself and we go into this in Chapter 7 but it is unlikely that you will need to do this in your first year – but in your second year, be prepared for all eventualities.

Conclusion

I hope that by now, you are getting a good idea of the little complexities of managing your bee hives. Nothing is difficult but some of it needs thinking about and as you progress you will gradually begin to understand just what an amazingly complex and fascinating society that you are dealing with. The next chapter aims to extend your knowledge by looking at a few of the problems that can occur so that you are fully equipped with the knowledge to manage your colonies during your first year and beyond.

7

Summer – Hive Problems and the Upcoming Harvest

Honey has a fuel value of about 3,307 cal/kg (about 1520 cal/lb). It readily picks up moisture from the air and is consequently used as a moistening agent for tobacco and in baking.

This chapter looks at some more beekeeping moves that you might need to know and some of the problems that might affect you during your first year with your bees. We are moving through the summer now and slowly approaching the harvest. Basically, if you have taken good initial advice and purchased wisely in terms of both bees and equipment there shouldn't be too much to go wrong but like any livestock, bees can be surprising creatures and you never know what's going to happen at times. We also talk about the upcoming harvest which is the subject of Chapter 8.

Regular inspections

Before we look at these other matters though, do remember that you should keep up a regular regime of hive inspections. You know now that these need to be more frequent during the swarming season and you should now have a good idea of how to go about this and what to do if you notice something amiss, such as no queen or no eggs, lack of stores or the existence of queen cells and so on. As the year progresses into summer, you can reduce the frequency of full inspections as long as you make sure that the bees have sufficient room to store honey for your harvest. One way of reducing the need for intrusive inspections after the swarming period (March to early June on average) is over is to use the external inspection routine.

This type of inspection can be used regularly and is a very effective way of ensuring that your bees are safe and well but can't replace an internal inspection. The main reason for this is because of the wide range of once exotic diseases that can now afflict bees in the UK. As I mentioned before in this book, it is no good claiming that you didn't know about the first stages of AFB (that cannot be determined by an external inspection) that have now become serious, because the government beekeeping inspector will still burn your hives and your neighbours will not be happy. There are also other reasons. I am sure that an inspection in good weather does not do very much harm to bees. I say this from the experience of inspecting thousands (literally) of hives which didn't suffer at all as far as I could see.

The external inspection

An external inspection can still tell you a lot about your bees and can help to reduce intrusive inspections. I mentioned before that there is a very good book on the subject by H. Storch (details in the Appendix of this book). It is now a bit dated but has some useful background information. Unfortunately, though not for him, many of the diseases now prevalent were not around then such as varroa and its complications.

Here is what to look for:

- *Bees not flying from one hive but flying busily from others on a fine day.*

This is an obvious one and needs a quick check to see if the bees in the non-flying hive are alive and well – or if there is a problem.

You should carry out this quick scan of the hives at any time you pass them – on the way to work/school/church etc and note any problem for later action. One good way of checking this is to put your ear up against the hive wall and giving a sharp rap on the side of the hive. If colony bees are present you will hear a sudden roar from within. If you hear nothing, expect the worst. You should now open up and check it out.

- *Many bees fighting at the entrance.*

This can be quite a dramatic sight and it is very obvious that the bees are fighting. It means that bees from another hive are trying to rob the hive you are looking at of their stores of honey. See below for further information on robbing and what to do about it.

- *Piles of dead bees at the entrance to the hive.*

This can indicate spray poisoning. Your foragers have returned to the hive and have been refused entry and have died. See below how to deal with this problem.

- *Dead drones at the entrance with other drones being dragged out.*

Drones are designed for mating. During a period of dearth such as winter or summer drought when there is little forage to keep the colony alive, other priorities such as colony survival come first and extra mouths in the form of drones are expelled. There is no need for mating at these times and so they are superfluous to requirements.

- *Mummified larvae littering the hive entrance and alighting board.*

This is a sign that the colony has chalk brood. The mummies are usually hard and of hexagonal shape. We look into this disease in Chapter 9.

- *Faeces spotting around the hive entrance.*

This indicates that the colony has dysentery. It can be quite dramatic in bad cases with much of the front of the hive covered. Chapter 9 looks at this in more detail.

- *Dead larvae being carried out but not carried away and often having been sucked out.*

This can indicate that the colony is starving especially if the larvae are being eaten as well. You need to feed the bees and this is covered later in the chapter.

- *A column of ants entering and leaving the hive. Also wasps entering. No guard bees present.*

This can indicate that the hive or nuc is empty, or all the bees have died. Carry out the listening and rap test to see what is going on in the hive.

- *Bees on the alighting board unable to fly even if prodded and moving in a moribund way.*

This can indicate a virus disease is present. I have seen this in one of my hives and the bee prodding test is a good indication. It can also indicate starvation and so a colony internal check for stores is recommended.

● *Strong, foul smell coming from the hive.*

This indicates that foul brood may be present. You should immediately seek the advice of an experienced beekeeper who will help you carry out an internal inspection. Don't delay on this in the hope it will go away. It could mean the destruction of your hives by the authorities.

● *Many bees flying at the entrance, usually in the afternoon but not always, facing the hive and appear to be bobbing up and down.*

This is a good sign and indicates young bees making first orientation flights. See robbing.

● *Many drones flying.*

This is normal for a colony on a sunny afternoon in the spring/summer.

● *A swirling, ascending mass of bees leaving the hive at a rush.*

A swarm leaving. Prepare to follow them until they hang up and then capture them.

● *Bees fanning at the entrance.*

There are two reasons for this. Firstly, they may be telling other bees that this is the home and they should come home. You should see the nasonov gland exposed when the bees are signalling this information. Secondly, the bees you see may be the last in a line of bees fanning to cool the hive. Bees will forage for water and spread it out in the hive and then lines of bees will fan their wings thus providing a refrigeration effect during hot weather. This shows an uncanny knowledge of nest thermoregulation and is to my mind one of the most fascinating things that bees have accomplished. It is a good sign, so don't worry about this and don't interrupt them.

● *Pollen entering the hive.*

This a very good sign and usually indicates that all is well with the colony and that the queen is present and laying.

There may be other signs that you will pick up on as you get more experienced. I know a commercial beekeeper who when entering one of his apiaries seems to know instinctively when one of his hives is having a problem and can scan an apiary of 60 hives in a few seconds. He will then point out a hive and say, 'that one has problems.' And he is right. These inspections are very useful but I say again, they are not a complete substitute for an internal inspection whatever anyone says.

Robbing

I mentioned robber bees earlier and this is a problem that can strike at any time. Bees will rob each other. It is a problem because bees usually attack other colonies that are weak and cannot defend themselves or small colonies in nucleus hives which are not yet big enough to withstand an attack. The best way to protect the small nuc from robbing is to ensure that the nucleus has plenty of ventilation from large openings covered in bee proof gauze (nuc colonies can abscond if they overheat) with a small one bee sized entrance that can be easily defended.

Colonies in hives that are being robbed pose further questions. Why are they so weak? Is it because of disease? If it is disease, are the robbers now spreading it back to their own hive? Yes, they are and so after you have managed to stop the robbing (if you can), you need to inspect the robbed hive the next day after everything has calmed down and find out what has happened and why. Seek advice because the problem may not be obvious. When I had just two hives (WBCs) in my first year of beekeeping, they both attacked each other! I have no idea why but there were fighting bees at both hives and all over the garden. My neighbours were becoming alarmed (and so was I) I took a hose to the lot of them and amazingly enough it all stopped. Until I turned the hose off and it all began again. The bees hid behind leaves on two apple trees when the hose was on. It kept going until dark when thankfully everyone went home and it didn't start the next day. I still can't think what that was all about but it was unnerving at the time.

In the meantime though, you need to see what you can do about the robbing.

Firstly, note which hive is being robbed and make sure that all small entrances such as cracks or badly fitting hive bodies and so on are blocked up with mud or grass or whatever comes to hand. Then reduce the main entrance to just a bee's width. Use a piece of wood, or grass or whatever. This will allow the bees being robbed to better defend themselves. If you place a large board

Robbing can get out of hand in an apiary and is difficult to stop.

across the entrance (not blocking it) this confuses the invaders as well and so gives even better protection. If the robbing gets out of hand you can as an extreme measure swop the two hives around. I have seen this done and it seems to work well but if the hive being robbed has disease, don't do this.

Certain strains of bee are more prone to robbing than other bees and Italian bees seem to be the first guys in. This may be because they are not thrifty with their stores.

It is unlikely that your bees will become aggressive in your first year especially if you have purchased a nucleus hive because small colonies are usually not aggressive but if you purchased full sized colonies then it is a possibility that they may be aggressive at times or even all of the time.

When I say 'aggressive', I mean more aggressive than usual. Most colonies will guard their nest even when smoked. That is why they have guard bees but sometimes you may do something that makes your colony more aggressive than usual. These actions include opening up the hive in wet or cold weather, especially thundery weather, or if they have been knocked over. Occasionally you do get colonies that are just plain nasty all of the time and every time you go near them it becomes an unpleasant experience. This needn't be the case and there are ways to deal with this.

Dealing with aggressive colonies

Firstly, see if there is anything that is causing them to be permanently aggressive. This could be something as seemingly innocent as a tree dripping on the hive during wet weather. This irritates bees and can cause aggression. Bees in the countryside can easily be upset by cows scratching their backs against a hive. I have had hives knocked over by this when my electric wire around the hives was broken by the cows in their eagerness to get at the nice grass inside the wire because they had nothing outside the wire! Bees situated under high tension wires can become aggressive. I have heard that this is the case but have never experienced it myself. Seek advice from an experienced beekeeper if aggression comes on suddenly and try and find out what could be the cause.

If it seems that it is a problem of genetics, i.e. it's the bees and nothing else, then your best bet is to ask advice from your association about re-queening your colony with a known gentle queen. This would be a queen purchased from a reputable queen breeder or a member of your BKA and inserted into the hive. The temper of the queen will translate into a gentle colony within a few weeks and your problem should be solved. Re-queening is not always a straightforward business and if you mess it up you could lose an expensive queen and still be left with an aggressive colony. Seek advice on this one. As I mentioned, it shouldn't occur in your first season.

A new queen?

What happens when you want to replace a queen for whatever reason? You have a nasty colony and need to change the genetics of the colony or because you accidentally kill her during an inspection or she just dies naturally and you want a queen of known provenance and gentle behaviour. The best way is to re-queen the colony yourself and this can be tricky. If you want to introduce a new queen into a colony you must kill the old queen first if she is not already dead. Even then, bees won't always take to a new queen so you must encourage them by firstly leaving them queenless for a few days. After three days look in the hive and kill off any queen cells that the bees may have started. (They work fast.) Then, having purchased a new queen and having received it in a travel/introduction cage, remove the plastic capping over the feed hole which is full of candy and place the cage between two brood frames with the entrance facing upwards. The bees will eat away the candy plug in the entrance and by the time they do this the queen will have (hopefully) been accepted by the bees and will be released. Once you have placed the queen in her cage in the colony, leave well alone for two weeks before taking a look. You should see eggs and young larvae. If so, all is well and you have re-queened your colony. Within a few weeks you should notice a difference in behaviour as your bees are renewed. In fact research shows that even the bad old bees change their ways with a new gentle queen. Re-queening is not difficult if care is taken with the process.

There are other ways of introducing queens and of introducing queens that you have reared yourself but that is for next year after further reading.

The laying worker

If you did lose a queen for some reason – and this often involves a beekeeper accidentally killing or injuring her during an inspection, the bees will generally use a young larva and create another one. These emergency queens are not always the best and at certain times of the year when there are few if any drones around, a poor mating may also result. For one reason or another, a queenless colony may be the result and the bees may not be in a position to replace her. The colony then becomes permanently queenless and this is one of the checks during your inspections. If this situation occurs, the absence of queen pheromones will trigger a change in some of the worker bees. Their vestigial ovarioles will grow and they will lay eggs. The workers then revert back to the females that they once were and will competitively lay eggs in the cells. These eggs are laid on the cell walls rather than the base of the cell because the workers are shorter than the queen.

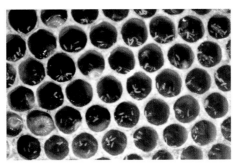

Multiple eggs in each cell spells disaster for the colony.

Several eggs will be laid in the same cells by competing workers. These eggs will result in drones (because the workers cannot mate) and these drones will not function properly. The visual signs are obvious and the colony is doomed unless early and urgent action is taken. You learned earlier how to unite a weak colony and a strong colony, well in this case you can't do that. The workers will probably kill the queen of the good colony and will not allow any others. If the colony is very weak and there are few bees left, then just disband it by taking the colony well away from the apiary and shaking the few bees out onto the ground. If it is still strong in bee numbers and you have sought advice and the colony looks salvable you can try the following:

Take the brood box about 200 metres away and remove each brood frame. Sharply tap each frame on the ground to dislodge all the bees and replace them in the box. Then return the box to the hive. The forager, non-egg laying bees on the frames will fly back to the hive. The laying workers have never been out of the hive and will not be able to find it. You must now either re-queen the colony or after a day, unite it with another hive. Again, you need expert instruction on this or if you are really on your own, buy a more advanced manual such as the *Beekeepers Field Guide* which will take you step by step through all beekeeping manipulations.

Spray poisoning and moving hives

This information is aimed more at beekeepers with hives in rural settings but as parks and gardens in cities are also affected it may concern urban beekeepers as well. As a new (small scale) livestock farmer, you will gradually gain an appreciation of farming needs and pressures and you will also begin to learn about what other farmers are growing because what they grow can feed your bees and give you honey. One farming requirement is to spray their crops with both herbicides (to kill competing weeds) and insecticides to kill predatory insects.

Your local beekeeping association will operate a spray protection scheme whereby responsible farmers will let the association know when and where they will be spraying and what with, and the association will let the beekeepers in the affected area know. When you know that this will affect your bees you can either close them up for the day or move them out of range for the duration of the spray. Just beware that you don't move then to another area being sprayed so check first.

Moving bees for short and long journeys

Closing up a populous beehive for a day or two in summer is a stressful business for the bees and although I have done it, if I just had two hives, I would move them. It's not difficult and your association will let you know where you can move them to. Try and move them at least three miles. If you move them closer, many of your foragers will return to their old site where you do not want them to be and they won't find the new hive site. I admit that moving a WBC hive is a bit more fraught because of all the parts so when I had to move my two WBC hives I removed the outer covers, placed a sealed crown board on top and tapped it with panel pins and strapped up the inner boxes. I also sealed up the hive entrance with sponge and placed both hives on the back seat of the car. I only had to move the bees a few miles from one side of Lincoln to the other and it took about half an hour so didn't allow the bees to overheat or become too stressed up. If you have a trailer then strap your hives to this, plug the entrances and move. If you move at night or in terrible weather there is not even a need to plug the entrances because the bees won't go anywhere. As a commercial beekeeper we moved thousands of hives by truck, always at night by strapping the stacked hives to the flatbed. We also were able to move bees in this manner during the day, if the weather was terrible.

So if you need to take your bees on a longer journey; e.g. to a heather moor, then my advice is to take them at night and to leave the hive entrance open.

But what if you need to move the hive just across the allotment or garden? This is a more likely scenario for you but also far more difficult and can be done in three ways. Firstly, move the hives a foot at a time over several days until they are in the required position. Moving the hives only a foot will ensure that the foragers are able to find their hives. This is a bit laborious and could take a while depending on distance. Another method is to just move the hive at night to the new position and plug the entrance with grass. By the time the bees have chewed their way out they will hopefully have realised that they need to re-orientate. Many bees however will still return to the old site so at this site place a box with comb (with stand and lid). Each evening for two or three evenings you can collect the combs of bees and place them in the hive on the new site. This is also time consuming and laborious. Both methods work but unless you really have to, don't move bees short distances. It is always best to move your bees at night. And a third method is to move the bees to a temporary site over two miles away, leave them there for a week and then bring them back to their new site a few yards from their original position.

Conclusion

This chapter should have equipped you with the knowledge to manage your bees through the summer period and face some of the problems that you may confront. Most problems can be dealt with if you think logically about them and if you have kept your inspection schedule up to date including frequent external inspections, you should be well able to keep on top of managing your hives. The next main task for you is going to be the harvest which can be one of the most enjoyable times in your beekeeping career if well planned and executed and the next chapter is going to tell you exactly how to go about this.

8
Summer into Autumn – Harvest Day and Post- Harvest Work

In preparing for battle I have always found that plans are useless, but planning is indispensable.

Dwight D. Eisenhower

I didn't realise that Eisenhower was a beekeeper but he must have been because he was obviously talking about the honey harvest when he said that. The honey harvest for the hobbyist is fun but planning is essential. You can involve your friends, your family and members of the beekeeping association and make a day of it with a barbeque and drinks to round off. Then provide everyone with a jar of honey for their help and you will have accomplished one of your aims in beekeeping in a fun and enjoyable way. For me, this is one of the most enjoyable days of the beekeeping year.

Planning is everything

The run up to harvest though does require planning and thought and to make a success of it you should prepare everything in advance so that on the day, you are not carrying dripping frames of honey into your kitchen only to find you have forgotten the extractor. There are several phases of preparation and they are as follows:

Your extraction plant!

This needs to be organised in advance. Firstly you need a bee-proof room. This is usually the kitchen but could be anywhere as long as you have hot water available. I extracted honey in someone's new kitchen once and it wasn't bee-proof. There were previously unnoticed gaps around the windows. It became impossible in the end as there were so many bees in the place. You couldn't move without getting stung. In the end we wore bee suits with gloves to finish the extraction and nearly boiled to death ourselves.

What equipment do I need?

You will need the following equipment:

An extractor: most beginner beekeepers won't have one but they are indispensable. Most associations have one for hire or loan and many of the local beekeepers may have one. If you are borrowing from the association, let them know well in advance so that you can borrow on the date you want. There are two basic types, the radial and the tangential. Both are fine except that with the tangential type you have to turn the frames around to ensure that they are not damaged and that all the honey is extracted. They remove more honey per pound of sweat than the radial but do have this limitation. Both types come as electric or hand models and I can assure you that it is far better to have an electric model. When using it, lift it up off the ground as far as possible or place it on a small table so that you can fit honey buckets under the honey tap. Extractors don't hold much honey before requiring emptying and if it is easy to empty then the whole process becomes easier.

Having organised use of a room and the extractor, you can now plan the rest of the tasks set out below but as we are dealing with equipment, we might as well finish the list of requirements.

A hive tool: for separating the frames, many of which will be stuck together.

A typical small scale electric tangential extractor.

Hand extracting keeps you fit.

An uncapping knife or *fork*: use either a special one or a serrated bread knife. If you are just using a kitchen knife, get two. One can sit in hot water while you use the other. Then you swap them. Knives are probably most useful, but the forks can be used on the awkward areas. Some beekeepers use steam heated or electrically-heated knives and these are wonderful. They can be purchased at vast expense from any beekeeping equipment supplier.

Honey storage containers: estimate the number required and try and get ones that fit under the tap of the extractor. I use those food grade plastic buckets that have tight fitting lids. They are big at 25 kg and will keep your honey safe.

A filter of some kind: I currently use a fairly wide, large domed kitchen sieve. I hang this from the tap of the extractor and the honey flows through this into the storage buckets. There will be lots of wax particles in the honey and these tend to clog up sieves so you may have to stir the honey in the sieve to keep it moving and clean it every now and then.

If you want really clear honey without any particles, for example, for show purposes, then you can line the sieve with muslin. This makes for really clear honey but it is a very slow business. Basically, the finer the filter, the slower the operation.

A muslin filter provides clean, clear honey but can be slow.

Commercial outfits use very fine filters and warmed honey is passed through them under pressure. Just about everything possible is removed and a sparkling but more sterile honey is the result.

Floor trays or *newspaper*: to stand boxes or frames on to collect drips of honey. Honey will drip everywhere. Believe me, you should place your boxes, both full and empty on newspaper or trays if you want it kept off your floors.

A large flattish metal uncapping tray: you will need a flat surface to uncap your frames on and this tray will catch the cappings and dripping honey. If the cappings build up on this tray, tip them into the large container (see below).

A large container: such as a bucket to hold the cappings that you have scraped off the frames. This needs to be quite large. At the end of the day, you will find that these cappings hold quite a lot of honey and it is worth hanging the cappings up in muslin bags (over a honey bucket) and letting it drip through. Some extractors have another device which can be fitted in place of the extractor basket. This cappings drum will spin and extract the honey from your cappings. The resulting wax is often very pure and yellow and many beekeepers melt this down for show purposes. Some beekeeping supply companies will swap wax for foundation so it is worth keeping all of your wax.

An empty hive box: to take extracted frames.

Plenty of wet cloths for cleaning: honey gets everywhere when you are extracting and so it is worth keeping plenty of clean, wet cloths around. When honey gets on the kitchen tap, bees will get on the tap and when you go next to turn the tap on you will be stung. I've seen and done it.

Preparing the hives for harvesting

Over the year to date you have regularly inspected your hives for stores and added honey supers when required. This year, you have left any spring honey surplus on the hives and have waited until the late summer and you have good supers of honey plus whatever is in the brood box. Now you must look in the hive and work out exactly what you are going to take from the bees and what you are going to leave for them. It is essential that you do leave them sufficient for the winter months unless you decide you are going to regularly feed them and that is not a good idea. When looking at your supers of honey, take a look at all the frames just to make sure that each frame has at least 75% of its cells capped with wax. The bees cap the cells when they have evaporated sufficient water from the nectar. If you extract too much honey from uncapped cells, the honey will in all likelihood ferment and blow up your jars, a bit like home-made wine always seems to – and it tastes terrible anyway.

If you need to make up full boxes with fully capped frames, just swap them around as required and this is acceptable as long as you leave sufficient space for the bees. So prepare each of your hives ensuring that the boxes with full frames of honey are available and preferably on the top of the stack. This can all be done three days ahead of H Day or even the weekend before if you are working on a weekend cycle.

Hard working slaves

Try and involve as many people as possible in this, especially family members, for two reasons. Firstly, because it is tiring, hot and sticky work spinning the honey and uncapping the frames, especially if using a hand-powered extractor and hot bread knives. With more people around you can work in shifts and have the odd refreshing drink in the off shift. Secondly, if you involve the family fully, you won't be getting in their way in the kitchen because they will be a part of the disturbance. This latter reason is quite important I've found especially when dealing with teenagers. No more irritable cries of 'have you finished in there yet Dad?' In fact people will actively be trying to slope off and avoid you, knowing that as official helpers, if they come near the kitchen they will be asked to actually do something! Actually it's also a lot of fun for everyone and your family will really enjoy it.

How to set up your kitchen

This factor needs a bit of planning as well. Firstly, you will need to place newspapers or trays on the floor where you are going to stack your boxes. Then you will need a table where you will uncap the frames over a large metal tray. Next to you on the table will be your bowl of hot water in which your two serrated-edged bread knives are kept or your steam plant or electric socket for your heated knives.

Once each frame is uncapped you place them in the extractor so this should be nearby on its stand and after the frames have been extracted, you place them in another set of empty boxes on the other side of the table. These boxes are also set on newspaper or trays because despite the fact that they have been extracted, the frames will still drip honey.

Now that you have planned the logistics of the day, you are well placed to begin work outside in the apiary and that always starts with parting the bees from their honey, and depending on how you do this, you can start the whole process either one or two days ahead of H Day, or on the day itself. Here is how to go about this.

Clearing the bees

In advance of the harvest you have to decide just how you will clear your frames of honey of bees. There are several methods. The nice way is to place what is known as an escape board with a bee valve underneath the lowest of your boxes that you are going to harvest. Do this the night before (or two nights before) H day. The bee valve allows bees to move downwards but prevents them from returning up into the supers. They work well especially in cool weather when bees go down to the brood chamber overnight to help warm the brood. They are not so good during warm weather. However, they do have a good effect overall. Use a board with a multi-valve array such as the Canadian escape board. This allows many bees at a time to move down and helps overall. Do not use the Porter bee escape. This hideous little device has a single valve based on spring wires. The wires soon get clogged up with propolis and don't work or drone bees get stuck in them so blocking all the other bees from going down. Knowing the system in UK beekeeping circles, you probably will use one but try not to.

Knocking/brushing them off

Yet another method is to clear bees frame by frame and for just a couple of hives, this is the method I often use. I open the top honey box up, lift out the frame one by one and sharply jar them over the hive. The bees fall off the frame and I then place it in a spare box down on the ground. I repeat this process until I have emptied all the boxes required. This works well but does upset the bees more than the other methods. I had my bees next to some horses once and harvest day became nightmare day. I have also tried brushing the bees off the frames and you can buy soft brushes for this but I found that they became sticky with honey very quickly and upset the bees just as much as jarring the bees. However, other beekeepers have had better luck than me using a brush claiming no significant build up of honey. Obviously it is better, the higher percentage of capped honey.

Moving the honey boxes to the kitchen

So far, you have cleared your bees and you now have boxes of honey comb ready to transport to the kitchen. As you work, keep them covered as much as possible otherwise you will have to clear them of bees all over again. Place a lid on the boxes after clearing. Now move them to the well prepared (see above) kitchen or shed or wherever you are going to extract the honey. How are you going to do this? If it is a car journey, cover the seat in newspaper or a plastic sheet. If you are carrying them down from the roof, just be careful. Box carriers can be used and are a useful item to have around as are sack barrows. However you move the boxes, do it quickly to avoid the bees finding their frames again. Once you have moved them to the kitchen or wherever you are extracting, you have now merged the two parts of your planning into one. You have your honey boxes in your kitchen and are ready to extract.

The extraction process

Just before you begin the extraction process, make sure your extraction room is fully secured and free of bee-sized gaps and you are ready to start.

Uncapping with a warmed bread knife.

This is easy but can be tedious. Simply take the first box from your stack of full boxes and place it beside you on the table. Remove each frame one by one and uncap, using the hot knife. Hold the frame on end in your large metal uncapping tray and slicing up or down, remove the capping wax. This will fall away into the tray. Use a fork to uncap those awkward bits where the comb is uneven and when all the cells have been uncapped, place the frame into the extractor. Repeat the process until the extractor is full – which is usually two to four frames for hobby extractors. Try and make sure that the frames are of equal weight otherwise the extractor, especially if it's an electric model, will come alive and rumble around all over the kitchen floor.

Spinning

If you use an electric extractor, the speed can damage the comb if you are not careful and with tangential extractors you should give a spin in one direction, then turn the frames round and spin fast and then turn again and again spinning fast. This helps to preserve the combs in one piece. Radial extractors won't require this change over but generally extract less of the honey. Hand extractors are more gentle and it is likely to be you that's damaged rather than the combs.

Once you have spun out the frames, place them into your empty box and once this box is full, place it on the newspapers or tray on the floor (honey will still be dripping). The box that the combs came from will now be your next empty box as you repeat the above process. Obviously, some combs are going to be really difficult to uncap but here, you have to do your best.

Keep a careful eye on the extractor while you are doing this. The level of honey will rise surprisingly swiftly and very soon, the extractor will be trying to spin through honey. This can blow the motor or you if you use a hand extractor, it can quickly exhaust you. Whenever this occurs, empty it straight through your filter into one of your storage buckets and resume operations.

If, right at the end of the spinning process you find that you have two frames left and you have a three frame extractor, then use the heaviest empty frame so that you can to get some balance in the machine, otherwise it simply won't work properly.

Cappings

You will build up a bucket of cappings and you will be surprised at just how much honey is mixed up in it. Pour any that is settling on top through the filter and then decide what you are going to do with the rest. There are three main methods of dealing with them that I have used and if you are careful they all work.

Leave them to drip

Hanging your cappings up in a muslin bag is a slow way to retrieve the honey in the cappings and the bag should be left hanging overnight at least. Good, clear honey will slowly drip into your container. The only downside to using this method is that the cappings will still be left with a lot of honey in them.

The oven method

Place all the cappings into a high-sided metal baking tray and place in the oven at a low heat (about 45°C). Leave for about an hour. Keep checking this. The wax will gently soften and melt and rise to the top with the honey underneath. Then pull out of the oven and leave to cool. It is a simple matter then to lift off the wax block and pour away the honey into a storage container. Don't try and speed it up by having the oven at a high heat as this can be dangerous.

The cappings drum method

For this, you obviously need a cappings drum fitting. When I purchased my first electric extractor, it came with this option. The idea is that you remove the frame holders from the extractor and replace them with the drum. This drum has fine holes in it and a solid floor so containing the cappings. Fill it with cappings, spin at high speed and much more honey will be extracted. The resulting dry cappings are usually of fine yellow wax and this can be melted down into blocks for later show purposes and we'll look at this later in the book.

And that is that. Extraction finished and clean up the mess time starts followed by a well deserved harvest celebration and a good barbeque.

Important note

It is important to know that two common UK honeys, oil seed rape (OSR) and heather honey are more difficult at times to extract. OSR honey granulates very quickly and so must be extracted very quickly, and if your bees have foraged from this crop you must keep a keen eye on the honey stores. This is possibly the only exception to the rule of leaving honey until you can extract in one go. Heather honey is what is called thixotropic which means that it sets to a jelly like consistency. If stirred, it will liquefy but then if left will again turn

A proud first harvest.

to a jelly. Special equipment is used to extract this delicious honey and so before you let your bees loose on the heather moors obtain advice from your BKA.

Storing honey

Most honeys will set in time and some will set hard so it is in your interest to decant the honey in your honey buckets into jars almost straight away. These can then be handed out to helpers, friends and so on without later having to break spoons trying to scoop honey out of a solid bucket of the stuff. Make sure that the jars are perfectly clean and that the lids and seals are also clean before decanting and you should have some good presents to give away.

If you have a lot of honey and you know that you aren't going to get through it all, it can be used for other purposes such as beer and wine making, mead making, baking and a host of other uses.

Selling honey

If you decide to sell honey at the local farmers market or at your garden gate, make sure you know the food production and sale regulations inside out. If

Selling your honey should not be difficult.

Figure 9: Honey labels need not be elaborate. Two of my early honey labels to fit hexagonal jars. Simple, easy to produce and very effective.

you don't follow them to the letter and someone complains about your honey, you will be in trouble. As new regulations come into force all the time, especially in the EU, there is no point in summarising them here because they will be out of date five minutes later but you will be able to obtain them on the internet or from the local authorities. These will include weights and how

Wooden crates and hexagonal jars: A point of difference.

to portray them on a label as well as other labelling requirements. My advice is that unless you have a lot of honey, place a nice label on a nice little hexagonal jar and just use the honey as gifts. Labels can be fun to make and a job for the kids. Above are shown a couple of my labels when I first went commercial in Spain. They were very simple and stressed the local origin of the honey. I sold it in small hexagonal jars in little wooden crates and was selling a quarter of a kilo for more than my competitors were selling kilos in large pots. Theirs was just honey. Mine was a souvenir as well as a local gift for folks back home or friends and family. Sell less and make more was my motto!

Testing honey

Honey with too much water in it can ferment and blow your containers so be careful about this. The limit is around 17–18% water. Anything more than that and you could have problems (unless its heather honey which can go up to 23%). So how can you tell if your honey is okay? Three methods to test it are as follows:

Rely on the bees. The bees only cap honey once it is properly mature so if you have ensured that 75% or more of all frames contain capped honey, then you should be fine.

Hygrometer. This can be used to check for water content but must be used while the honey is still liquid so do it immediately after extraction.

A honey refractometer. This is a more specialised device and there are two types. The optical device needs proper calibrating and is simple and accurate to use as long as it is kept calibrated. The electronic device is even simpler and gives a digital reading of the water content. Simple but expensive!

At first, I relied on the bees, and even after I became a commercial beekeeper I still used this method for years. I eventually bought an instrument because honey buyers wanted to know the water content. My advice for your first season is 'trust the bees'.

What about the wet frames?

Obviously you will have a number of wet frames in your box stack in the kitchen and you are going to have to clean them up. If you simply store them, you will have all sorts of minor beasts attacking them especially ants, wasps and others. In my view, the best way to clean them of the rest of the honey is to place the boxes on end with the frames free to air in the apiary. You had better not do this if your hives are sited in a small garden because there will be a lot of bees floating around and it would alarm your neighbours unnecessarily. If you have plenty of space however, you can do it this way. Many will caution against it saying that it will encourage robbing but I have never seen this in my years of beekeeping and I'm not sure why it would. Within a day, the frames will be dry and you can then store your boxes of empty comb (and we'll look at storage of comb later).

Another way is to simply place the boxes back onto the hives. The bees are then more contained and the frames will be cleaned by the hive bees.

Wax comb storage

If you need to store surplus honey boxes with comb over winter, then be careful. Wax moths love unguarded comb. Pure foundation wax is of little interest to them as it holds no nutrients and so it is safe, but comb, especially brood comb is a prime target for their little teeth. This makes it difficult to store both comb and cappings wax and even propolis which has much wax in it. I once sold a barrel of propolis to a buyer and when I opened the barrel to show them what fine propolis it was, to my horror there were a million wax moth larvae looking up at us out of the usual wax moth mess, but an angling shop took them in the end.

There is one good way of preventing wax moths from attacking your stored frames and that is to buy some Bacillus thuringiensus (usually goes under different trade names but bee supply shops will sell it). Mix it with water, spray the frames lightly with a small garden sprayer, dry them in the sun and store them safely. No wax moth larvae will touch them, or rather they will try but die in the attempt. The bacillus doesn't contaminate anything except the moth larvae. The risk to this is that if you leave the frames too wet and they contain pollen, this will get mouldy.

Some beekeepers will leave their boxes of empty comb on the hives over the winter period but there could be problems with this. If there are not many bees in the colony then they won't patrol any extra boxes and wax moth may gain a hold. It is what I normally do but only for a few boxes (one per hive) and only on strong hives.

You can also freeze the frames. This kills off eggs but then unless you keep them frozen, the problem will return. Never use PDB crystals (moth ball stuff). It has been listed as a carcinogen and contaminates the wax which is a chemical sponge.

And now, *after* the harvest, it's varroa time again.

Varroa again

Once your harvest is completed you should think about your autumn varroa treatment. Again, find out what the local association beekeepers are doing and when and follow suit. If you are going to use an organic treatment follow the instructions very carefully because as I mentioned earlier, they are not as easy to use as the chemical strip treatments due mainly to being temperature dependant. You should always treat the bees AFTER the honey has been removed which avoids any contamination. I would carry out this task in September and remove the treatments as instructed in the accompanying information leaflet.

Conclusion

This chapter has hopefully brought you through a trouble-free harvest and given you a good appreciation of post-harvest tasks as you move towards wintering your bees. Take a look at the annual bee calendar and checklist in the Appendices at the end of the book. This will provide you with a monthly action summary of all the jobs that a beekeeper must undertake during the year as well as showing you what the bees are up to at the time. Keep it with you and use it so that knowledge of what to do and when becomes automatic. Use it to learn about what your bees are up to.

Chapter 9 looks at how to take your bees through the winter period and into spring – and back to May/June time when you will have been a beekeeper for a year.

9
Winter and Beyond

*Secreted from glands, beeswax is used by the honeybee to build
honey comb. It is used by humans in drugs, cosmetics, artists'
materials, furniture polish and candles.*

The year past

It is now winter and your bees are hopefully alive and well and you are leaving them alone to come through the winter. During the year since May/June, when your bees first arrived, you have learned a huge amount of beekeeping knowledge and become confident when handling your bees. A summary of your learning up to now should go like this:

- What type of hive to use and where to locate it (them).
- How to receive and install bees in an apiary.
- How to inspect a colony of bees and know what to look for during the inspection.
- How to assess a colony of bees during an external inspection.
- How to manage your hives during the swarming season.
- How to catch and install a swarm of bees.
- How to deal with a variety of problems that can arise during the year.
- How to carry out various manipulations that may help your bees during the year
- How to prepare for, harvest and store your honey.
- How to manage essential post-harvest tasks.
- What to look out for when inspecting for disease including reporting obligations and who to seek advice from if you notice a problem.

This book has told you a lot about the subject but there is only so much a book can say. There is nothing like being out there in the apiary to give you real experience and a real learning experience. In this chapter I will remind you of the tasks necessary to take you through to May/June and the anniversary of where you started which means getting your bees through winter and into a new spring.

Moving towards winter and pre-winter tasks

Final inspection

As autumn progresses and winter draws near, the drone bees will mostly be expelled from the hive and the queen will slow down her egg laying. The colony will gradually be preparing for winter.

As the colony moves into winter, the beekeeper should make a final inspection of the beehives to ensure that there is no sign of disease and that there are sufficient stores to last through this period of dearth. Remember that even though there might be nice sunny and warm days during the winter when the bees will fly, there may be no source of nectar for them. While you are doing this inspection, it's worth cleaning the floor if you are using solid floors. Take it off and scrape any build up of debris that has collected. Remove the queen excluder and place a shallow or another brood box of honey on top so that you have the required amounts of feed. If winter doesn't begin to bite, and you can never be sure these days, then it is worth checking after a month to see if the bees have enough stores. As I mentioned earlier, in warm winters they may fly more often and may deplete their stores at a faster rate and so may need feeding. If however you have been generous in leaving your bees with ample honey, this should not be required.

Also check that there is a queen and that she is laying. There must also be sufficient bees as well – at least six frames-worth in the UK. More further north. The reason for this is that there must be sufficient bees to form a decent cluster for the purpose of keeping an area at the correct temperature for the queen and brood.

Colony stores

You should have left the bees with sufficient honey to last them over winter and the amount required assuming an average winter temperature of —4°C to +10°C is 15–30 kg. Each shallow British National frame holds 1.5 kg and each full frame holds about 2.5 kg. Langstroth frames hold about 2 kg and 3 kg respectively, so you should easily be able to work it out.

There are two basic feed mixes that I use during the year if required: a thick syrup for autumn feeding which will be stored more or less immediately, and a thin syrup for spring or stimulative feeding which is consumed without storing and is used in the very early spring. The mixes I use are as follows:

Thick – 2 lb (1 kg) sugar to 1 pint (½ Litre) of water.
Thin – 2 lb (1 kg) sugar to 2 pints (1 litre) of water.

There are many types of feeders available but I find the most convenient is a frame feeder. This is simply a container the same size and shape as a normal hive frame. I place it at one side of the box after removing the frame. If you use these feeders remember to place floating pieces of wood or dried bracken into the feeder so the bees don't drown. I have also used contact feeders which can be any container with a secure lid – jam jars, small bucket containers and so on. The lids of these containers are pierced with small holes and the container is upended over the feed hole of the crown board or directly onto the top of the frames of the brood box if no crown board is used. Because it can be made from any conveniently sized container, it is easy to make and costs little if anything. Place an empty box around the feeder and replace the hive lid.

Frame feeders are easier to use and convenient.

How much sugar to feed if necessary

To work out the amount of sugar syrup to use, remember that each 5 litres of heavy syrup will increase stores by 3 kg.

Remember that the colony needs pollen as well so check this by ensuring that there is a good arc of pollen on the brood frames. The bees need pollen especially for brood rearing and won't use much during the winter. Their main requirement is for early spring use to feed brood. If there is no pollen and there are few if any early pollen sources, then think about placing a pollen pattie over the top frames in the very early spring. These can be purchased from bee supply stores and are simple and easy to use.

Other wintering tasks

There are other tasks that you need to carry out after the harvest as winter approaches and these should all be complete by the end of October.

Check that the hive lids are sound and totally waterproof especially if you lift them slightly on matchsticks to aid ventilation. A mesh floor and a raised lid may sound as though the hive will get too cold but remember the bees are keeping the centre of the cluster warm, not the hive.

Check that all woodwork is sound and that there are no holes in it for wasps or other pests to enter.

Remove the queen excluder if you are using one and store it in the lid. This helps the winter cluster move around the honey stores.

Wind danger. You will know best the climate in your area and if you suffer from winter winds, then make sure your beehives are well strapped together and you have bricks on the roof.

Place a mouse guard over the entrance to the hive. You can buy these cheaply from any beekeeping supply store. This allows only insect-sized creatures to enter and prevents mice from nesting in the hive. They cause a lot of damage and are usually left alone by the bees which may be clustering.

Arrival of winter

Winter has arrived and your worker bees will not be doing as much work and so will live longer than their summer compatriots and your colony will cluster together once the temperature inside the hive gets low enough. This cluster is the honey bees' way of using their body heat to maintain an exact brood

Hives in the snow. Make sure the entrances are not covered.

temperature in the centre of the hive. This temperature keeps the brood safe if there is any and allows the queen to lay eggs towards the end of winter and so gain a good start on the year. The cluster loosens and tightens depending on the temperature and is another reminder of the bees' amazing powers of thermoregulation in the hive. Like termites, bees operate their nest heating and cooling regime with precision.

Keep an eye on your hives over winter. There should be no need to feed them and definitely no need to place candy on the frames. You should have ensured that the bees have enough stores to last them throughout the winter but if on a fine, warm, sunny winter's day your bees are flying from one hive and not the other, check it out. If they are starving, feed them and an emergency feed is just to put ordinary, dry sugar crystals on the crown board or frame tops. Not the best, but the best in an emergency when you haven't got candy or sugar syrup to hand. I have saved hives this way. Candy can be simply made from 2 kg of icing sugar mixed with 1/4 of a teaspoon of tartaric acid and 2 teaspoons of glycerine.

If your area receives a lot of snowfall, ensure that your hive entrances are not blocked by snow.

Checking colonies in winter

Over the winter period just use commonsense when dealing with your bees and leave them alone if possible. Winter should be a quiet time for your bees but keep an eye on them. Snow, rain, winds, flooding and so on are all risks that you must manage. Keep your colonies dry and as much out of the wind as possible. Take a look on fine warm days (and you can get them in winter) to see if your bees are flying and if they aren't, put your ear to the hive and give it a sharp rap. If there is a healthy roar from the bees, then all should be okay but if nothing happens, then check it out, quickly and smoothly.

Dead colonies

If the hive is dead, try and find out why with the help of a more experienced beekeeper in case it was a disease then clean everything up, dismantle it and store your boxes away. The colony could have died from starvation, disease that you hadn't noted earlier and so on, and it is well worth finding out why. If, for example, it was because of AFB, then all of your kit needs sterilising and it is a good idea anyway to sterilise any hive if you are unsure. The best way to do this is to use a blow torch. Just play the torch across all surfaces and especially into corners and cracks and you should end up with a clean hive ready for a new colony. Be careful about this, remember that wood and wax are very combustible. (Don't do this on hives made of polystyrene or plastic for obvious reasons.) Then think about what you have learned and change your game plan next year if necessary.

Winter thinking

If your hives go through the winter without any problems which is more likely, then you should be able to relax and think carefully about the beekeeping year ahead.

Should you expand?

If so, how many more hives? Remember all the new boxes and frames that you will need. Two extra hives doesn't sound much, but each hive will need two brood boxes and four honey boxes. Two hives will therefore need two extra queen excluders, 12 extra boxes, 120 extra frames and 120 sheets of foundation plus the room to store them all when not in use. Then there will be all the extra honey to deal with and 100,000 extra bees in the garden! Don't forget the neighbours.

Should you re-queen your colonies?

Remember that colonies with new, young queens swarm far less readily than those with older queens. If you are going to re-queen, should you use your own, reared queens? (If you go this route you will spend some of your spring doing this and it probably isn't worth it unless you have more hives because one hive at least will be used for queen rearing.)

What about the wax?

What will you do with all your wax that you have collected from the harvest and from scraping off burr comb during inspections? (We take a further look at this later in this chapter when we look at building a solar wax extractor.)

New purchases?

An electric uncapping knife? A proper honey filter? Obtain catalogues from the various beekeeping supply companies and look at them at your leisure over winter and then go and indulge your hobby. Think about buying some thin unwired foundation and a comb cutter so that you can produce cut comb honey next year as well as liquid gold.

If you look at your notes and talk to other beekeepers at association meetings, you will very quickly come up with new ideas for the year ahead that will help your beekeeping. For example, talk about smoker fuels. Something that won't go out every two minutes! The possibilities for new ideas and directions are endless and we will look at these later in this chapter.

Winter projects

There are several worthwhile winter projects that you will now have time for, such as perhaps new and better hive stands and so on, but one in particular is well worth mentioning and that is making a solar wax extractor.

Solar wax extractor

This is not only easy to build but is extremely useful. What will you do otherwise with all the wax you collect. The bees use a lot of energy making it and it is easily recyclable so why not do it? There are simple plans available on the internet (www.beesource.com) and if you want to tweak these a bit you should paint it black and double glaze your finished article. These devices work superbly even on days when you would expect little sun but they do get very hot inside so be careful. Basically you place old frames and pieces of wax into a sloping tray in the extractor. On a sunny day, the inside of the extractor heats up very swiftly and the comb and frames melt and run through a grid which retains all the bits and pieces. The wax drips into a removable container until full. Then the container is removed and left to cool over night. The resulting wax block shrinks slightly on drying and is easy to remove from the container. The purity of the wax that you extract is pretty good because any rubbish tends to stay in the combs and you should be able to render nice clean blocks of wax for resale or re-use. You can swap wax for foundation with many bee supply companies and so you get free foundation. Okay, the company isn't going to lose out on the exchange but it is an excellent way of keeping your costs down and it has to produce the foundation. You could also make your own foundation by purchasing a single sheet foundation maker – another idea for a present. Beeswax is a very much a multi-purpose product of the hive and has very many uses – polishes, creams, cosmetics, candles and many other items, and we will have a look at this aspect of beekeeping later on.

A four- or five-frame nuc

Another good winter project is to make up some nuc boxes for various purposes. You would need them for queen rearing, swarm catching and as you will learn in your future beekeeping career, many other purposes. They are not difficult to make and as you know if you purchased a nuc of bees to start in beekeeping, they are simply four or five frame boxes with an entrance hole and a lid. See: www.beesource.com again for the plans and get building.

If you look at much of the woodenware in the catalogues, there is little that cannot be built if you have a modicum of carpentry skills and a shed.

Books and magazines

It is a great time to look through specialist beekeeping books in the winter. There are many out there and it is worth looking at the reviews both in magazines but especially on Amazon where reviews from ordinary beekeepers are to be found in abundance. My recommendations for a variety of more specialist books are found in the bibliography and as for magazines, there aren't that many so ask for samples and see which ones you like. These magazines and books will show you all of the almost endless possibilities for study and interest open to you for the future.

And so through to spring and the need for pollen

The queen will begin to lay eggs in the warm cluster surprisingly early (some never stop) so don't be taken by surprise at the build up of bee numbers in very early spring. This is when both early nectar sources and early pollen sources are vital. Lack of pollen at this time of year can cause the colony to build up really slowly and some beekeepers may assume Nosema or acarine and so on and make the wrong moves. Make sure you see pollen entering the hive in abundance from crocus, rock rose, willow and so on. If there is no pollen you have to feed pollen patties, and as I mentioned before, these can be purchased from bee supply stores. Simply lay them above the frames of the brood box and the bees will eat them. Before doing this however, take advice on the matter from your beekeeping association.

Arrival of spring

Now with your new knowledge and the experience gained with nearly a year of beekeeping behind you, you will really enjoy the second year that much more. You won't be so much in the dark. You will have a much better idea of what you are doing with your bees and in which direction you want to go. You still need to get your bees through to May/June of the new year and with colonies that are now true colonies and not nucs or swarms, it becomes a bit trickier.

The exact date for this is going to vary year by year. Don't take any notice of official springtime dates. Everything depends on the weather. Bees depend on the weather so as the days gradually get longer and slightly warmer, the colony will react. The queen starts laying in the cluster sooner than you think, almost in advance of the better weather and this will be the basis of the phenomenal spring explosion in bee numbers.

In your second year now your bees are building up to swarming and you must go back to chapter eight and remind yourself of how to go about controlling this natural process. On a fine day in early spring (say late February/early March), remove mouse guards, place the bottom brood box on top of the top brood box (i.e. exchange their positions) – if you are using two boxes. The bees tend to move upwards and you will probably find that the bottom box is empty. Give the bees room to expand upwards by placing this empty box on top. Then replace the queen excluder and place a honey super with drawn comb if possible above this. The bees will now have room for the queen to lay and room to store early nectar and you will have given yourself time. Only do this on a fine day and do it quickly. Exposing brood to cool temperatures can be fatal. Remember, they need their 34°C to survive.

Check the stores and feed if necessary and on warm days check that the bees are flying and that pollen is going into the hive. It is easily seen as brightly

Pollen entering the hive is a welcome sign
and is indicative that all is well with the queen.

coloured little balls on their hind legs and it is worth repeating here that pollen
is as important as nectar for the colony.

Varroa again

Spring is also the time to treat your bees for varroa and this is very important.
Remember to seek advice about timing and type of treatment from your local
association and with your colonies bigger than last year, the incidence of
disease could be greater so make sure you keep a careful look out for any signs
of disease during your inspections. Hopefully, the local association will advise
alternating treatments to avoid a build up of resistance by the mite but if not,
why not ask about this?

Swarming

This is now really important. In this new spring, your hive will not be the small
nuc or swarm of last year and will therefore be that much bigger and will
expand that much faster and will be ready to swarm that much earlier and
more readily. Your queen will be older and in her second year will swarm more
readily than a younger queen and this swarming business will engage your
mind much more than when you started beekeeping. Now go back to the
information on swarm prevention in Chapter 6 and ensure that you use the
advice it contains. Keep an eye on things and act swiftly.

Keep on top of your bees now with regular inspections, and if you are really
keen to prevent swarming, you can re-queen your colonies or be pro-active and
perform an artificial swarm. Just make sure you have sufficient kit for this. My
advice during this period is to keep a good watch on your colonies and be
proactive.

There is also of course a benefit of swarming and that is that you could put out a bait hive or two to 'acquire' the stocks of those beekeepers who aren't as vigilant as you! They're free but remember to take a good look at them after you have installed them – and again remember to have the required kit immediately available.

Disease

Larger, rapidly building colonies can attract disease and you must keep an eye out for this. Chapter 10 gives you some hints and instructions on how to go about this and like swarming, this is really important now. Look out for an unexplained failure to build up – even though there is abundant forage. This could be a sign of a problem. Go to Chapter 10 and obtain advice from your association.

Harvesting advice for the new year

I alluded to this at the start of the book. One piece of advice which beekeepers sometimes forget about is to be careful how you look after your bees towards the end of spring. In many districts there is a dearth of flowering plants after spring and if you neglect to check the bees' stores over this period, your bees could starve. This takes many beekeepers by surprise. Some will see their hives bulging with delicious spring honey and without thinking about the local floral scene will harvest it believing that the bees can continue like this into the summer and beyond. Keep this in mind and always know what flowers exist in your area, their relative abundance and their flowering times. In some parts of Europe such as my area of Spain, there were two flowering times – spring and autumn with nothing in between and sometimes the autumn period was poor. I had to be very careful how much I harvested each year. Remember to learn about your honey flows.

Conclusion

At the end of this book there are Appendices covering many of the offshoots of beekeeping as well as further reading advice, hive products and other allied subjects. Do take time to read them because they will give you plenty of ideas for the future. If all is well you should now be ready to head into your second year of beekeeping with confidence and growing knowledge. You now have a huge amount of both theoretical and practical beekeeping knowledge. You can manage bees right through the year and keep them safe over winter and you now know where to go and who to ask if you find problems that you can't (yet) deal with. True, there is a lot more to beekeeping which is outside the scope of this basic primer but I have provided a list of useful books and recommendations in the Appendices covering advance beekeeping manipulations as well as other subjects and interests to do with beekeeping. As you have come this far I'm sure that you are on course for many years of learning about these essential insects. Good luck and good beekeeping. BUT before I finish, take a look at Chapter 10 which goes through diseases and pests and learn it well.

10
Bee Diseases and Pests

Great fleas have little fleas upon their backs to bite 'em,
And little fleas have lesser fleas, and so ad infinitum.
And the great fleas themselves, in turn, have greater fleas to go on,
While these again have greater still, and greater still, and so on.

Augustus de Morgan

For fleas, also read bees! This is really a huge subject and there are whole books written on the subject but the aim of this book is not to make you into an instant bee vet (you will become one over time), but to help you identify the fact that something in your colony is wrong. Diagnosing diseases can be a tricky business. Some are obvious and others all look the same and have the same signs and symptoms. The aim of this chapter is to guide you in noticing that a problem exists and then once you think there may be a problem, you can go and seek advice. There are two main types of disease: diseases that affect brood, and those that affect adult bees – and this chapter will give an oversight of both types.

How to tell if a colony has a disease

This is often difficult even for experts but in my view the best way to notice disease in a colony is to know what a healthy colony looks and smells like. Your new swarm or nucleus colony after say two or three weeks will be perfect and here is how to learn:

Go to the brood box and lift out a frame near the centre. One with the arcs of brood, honey and pollen.

Look at the single eggs sitting upright at the bottom of the brood cell.

Look at the coiled, pearly white larvae. Look at the very young larvae like little commas sitting in their pools of milky white royal jelly. Look at the capped larvae and note the even colour of the neat, slightly domed cappings. Note what they look like and remember it. Smell it (I mean it!). Know the smell of a healthy brood box and a healthy frame. It should smell fresh and waxy and honey-like and it is an unmistakeable smell. Even after just a few inspections you should know and remember this smell. If anything is different in look or smell, then seek advice.

If you hear that another beekeeper in your area has a problem with disease, ask if you can go round and see what it looks like. Look at how it differs from the healthy colony above. I did just that when a brave beekeeper in my local association in Lincoln rang everyone up and told us one of his colonies had AFB and we should get there quick to have a look before the bee inspector arrived. I saw the difference and the various symptoms and never forgot them.

What differences could there be and are they significant?

Differences to the above are often not difficult to spot but interpreting the differences can be tricky, so let's look at some examples of these and see how we can work out what is happening.

- Take a close look at the larvae both small ones in their royal jelly and the larger coiled ones. If they are off colour, i.e. not pearly white or milky white, then there may be a problem.
- If the larger larvae are not neatly coiled but a bit straightened out or have a melted down appearance, then there is a problem.
- If the wax domes covering sealed larvae are sunken or pitted or slightly holed. Then there is a problem.
- If there is a slightly bad or very bad smell coming from the brood frames or brood box then there is a problem.
- If there are whitish ribbons running through the wax comb resembling tunnels, then there is a problem.

All of these 'differences', except the last, are symptoms of a brood disease and is likely to be either American Foul Brood (AFB) or European Foul Brood (EFB). Both are dangerous especially AFB which is deadly and very infectious and remains so for years. Both are notifiable in that you must notify your local bee disease inspector that your hives have them. If you don't know who he or she is then contact your BKA. Having noticed the 'differences' you should then

Spot the difference! Healthy and unhealthy brood.

seek advice from the association or from the local bee inspector. They will help you and tell you exactly what to do for the future. The last 'difference' noticed above was evidence of wax moth larvae tunnelling through the combs.

These were just examples of the sort of things you should be looking at every time you carry out an inspection. It wasn't meant to be a definitive disease check of any kind but you should now see what I mean. You must remember though that some symptoms are very much like others and it can get very confusing. Also, very early symptoms of diseases such as AFB are difficult to see even for experienced beekeepers and so by the time you do notice something it can be quite advanced. All will come with experience though so just learn as much as you can as you go along. It's now worth going briefly through the various diseases so that you can get a broad idea of what can go wrong. For detailed explanations and diagnoses/treatments, it is best to read a more advanced book such as *The Beekeepers Field Guide*.

Pests and diseases

Wax moth

I've started here with wax moth which although not a disease it usually means that the colony is failing due to a disease. The part wax moths play in nature is important. They destroy ailing colonies. By doing this, any disease won't spread to other colonies. If you find this damage then look for signs of disease.

The moths come in two sizes, large and small, and essentially, any small, grey/brown, dull coloured moth in the hive will be one. Their larvae eat everything and leave a total mess of dirty webbing and frass and their cocoons bury themselves in the woodwork of the frames and hive walls damaging the

Typical wax moth damage. Look for the reason why the colony can't defend itself.

wood. Look out for the plump white larvae and the larval cocoons. A healthy colony will always keep them in check, so if you find damage, check it out and seek assistance.

Brood diseases

American Foul Brood (AFB)

Notifiable. Very infectious for years as the causative bacterium has a spore forming stage. Larvae usually die after being capped. The cell cappings are often punctured and discoloured and sunken. If the cells are open, at a certain stage you can prod the brown larva with a match and draw out a string of goo. If the larva has dried out it forms a hard to remove scale in the cell with the pupal tongue left protruding. This is a typical and very noticeable feature as is the foul smell. Early features show a brood frame with many open cells amongst sealed cells giving a pepper pot appearance. In the UK, only a regional bee inspector can sort it out and they will confirm the disease (or not) and take the appropriate action. If AFB is confirmed the inspector may order the hives to be burnt. This eradicates the problem and stops it spreading and this method has had huge success in the UK and New Zealand in minimising the problem. You must then wash all you clothes and gloves and clean all your kit. Any hive parts that survive the burning must be scorched with a blow torch. This disease is serious and if you fail to do something about it you will be in trouble with the authorities.

European Foul Brood (EFB)

Serious, and again notifiable, but the bacteria has no spore forming stage and so is far less infectious than AFB. The best way of noticing this disease is by seeing the larvae change from pearly white to off white/yellow in colour and you can notice the trachea (breathing tubes). Also, the larvae adopt unusual positions and often have a melted down appearance. They die of starvation because of the parasitic nature of the bacteria. The later stages smell really bad. Again only the bee inspector can decide what to do and often in bad cases burning is the answer. If the infection is light and the colony is strong, it may clear up on its own. I have seen this happen.

Chalk brood

This is a fungal disease and it affects sealed and unsealed brood and the larvae become grey and discoloured later becoming hard, white and shrinking to a small hexagonal shape. At this stage you can tip the comb on its side, tap it and the mummies will fall, out. This disease is probably the easiest to recognise and when the bees carry the chalky, white mummies out of the hive they often drop them on the alighting board. If you see this, you know your colony has chalk brood. There is no satisfactory 'cure' and even though I have had hives with chalk brood, it has usually cleared up of its own accord.

Sac brood

This is a viral disease and rarely causes severe losses. Affected larvae are often removed by workers and so if you notice it then it is fairly advanced. The larvae bloat and resemble bloated sacs. This is easily removed with a pin but if left it dries out to a scale resembling a long Chinese slipper with an upturned end. Colonies often recover spontaneously when a good flow starts but overwintering adults which carry the virus can start another outbreak in the spring.

Chilled brood

This isn't a disease but is common when a colony is depleted of nurse bees. This can happen when foragers are poisoned and are replaced by nurse bees.

The signs are obvious: dead larvae especially around the edges of the brood area. Another reason for this is if you move a brood frame away from other brood frames thus isolating it from the main brood area. Be careful when inspecting the brood box and keep your colonies strong.

Adult bee diseases

Nosema apis *and* Nosema ceranae

Two variants of a similar disease caused by a microsporidian (now considered to be a fungus) which has a resistant spore stage. It is common and widespread and if spores are eaten by bees, they germinate, invade the gut wall where they multiply and produce more spores which are passed out in the waste. This is often cleared up by other bees and so it goes on. It is common in spring and autumn and where permitted beekeepers use a substance added to sugar syrup called Fumadil B. This antibiotic inhibits the spores from reproducing but doesn't kill them. The failure to build up in the spring is a good sign of nosema but other causes must be investigated as well. Other than that the only way to tell is to grind up a dozen bees and look at the result under a low power microscope. If you see rice shaped organisms, they are likely to be nosema spores. (Or send some bees to the lab.)

Nosema ceranae

This is similar but can kill off a colony faster than *Nosema apis*. It was first discovered in Spain in 2004 and has the effect that the workers are too weak to return to the hive leading to rapid decline in colony strength and a Marie Celeste atmosphere in the hive. It is associated with colony collapse disorder (CCD, see below) and is thought to be one of the causes of this.

Good hive management and keeping colonies strong and stress-free is probably the only management practice that can keep it at bay.

Dysentery

This is usually caused by excessive water accumulation in the gut from eating contaminated stores. It is a symptom and so look at the stores to see what they are eating. It can spread nosema easily but is not caused by nosema as many texts claim. Keep it at bay by ensuring that any feed you give your bees is clean and pure, especially winter feeds. Signs are easy to spot and you will see faecal spotting on the hive and in severe cases it can cover the entrance to the hive. You will also see the same on the combs with nurse bees cleaning it up and ingesting nosema spores (if contained in the faecal spots) at the same time.

Virus diseases

This term covers a variety of problems caused by those ingenious half life things called viruses. You will hear many exotic sounding names such as Kashmir bee virus; Israeli acute paralysis virus; black queen cell virus; deformed wing virus and so on. There is no way a beekeeper can work out which one is which but many are endemic in the hive only surfacing sufficient to be a problem if something else goes wrong or the bees become stressed from other causes. One real friend of bee viruses is Varroa destructor (see below) which is thought to have the effect of puncturing the bees and vectoring in viruses that are already in the hive.

Signs of a problem are difficult to detect but moribund, dozy bees with a greasy, black appearance sitting on the hive or alighting board are a sign. I've seen this in one of my hives in Spain. I could prod the bees and generally push them around and they did nothing. They really looked black and greasy. It's not just a phrase. I didn't know what it was until a local expert told me it was 'The Bad Black' (El Mal Negro). I looked this up and found it was thought to be a virus disease. I did nothing (not the best way to do things) and it went but that hive didn't produce a surplus that year and if honey is your living, that is not good.

Colony collapse disorder (CCD)

This becomes apparent when there is an insufficient work force to maintain the brood that is present and the workforce that exists is made up of young adult bees only. The workforce seems to rapidly decline to the point of not being there anymore. I recall on one visit to my hives in Spain, the colonies seemed fine. Two weeks later nothing remained. No bees, no dead bees, just dead brood and amazingly, no sign of wax moth. This was in 2004 and I couldn't work out what had happened. I'm still not really any the wiser even though I have read everything there is to read about CCD. Perhaps it was Nosema ceranae which was found in Spain at the time. Other hives were okay. Now, there are whole research programmes into the causes of this because governments are getting truly worried. Try and keep up with the research and go to any lectures about the subject offered by your association and keep varroa at bay. It is a global problem.

Bee pests

There is always some sort of connection between varroa and CCD and bee viruses so don't think of them as something apart. Learn about all of them.

Varroa destructor

We have discussed this before and it caused probably the most significant change to beekeeping in most of our lifetimes. It is a mite that you can see both on capped larvae (if you open them up) and on adult bees. It causes problems to both and having come from a bee environment, it is perfectly adapted to live in beehives. It is associated with CCD. Varroa cannot be ignored. You must monitor it and treat it and the way you do this should be in coordination with other beekeepers for reasons that have been discussed earlier. Use your beekeeping association as a source of advice and ideas. This is the only effective way to combat varroa on a regional basis and ensure that treatment methods are the best available. By combating this mite effectively, beekeepers can I am sure help to lessen the incidence of CCD in our hives. This can only be a good thing.

Monitoring varroa

Monitoring varroa in your hives can be a useful exercise and I generally monitor hives four times per year, early spring, at the start of summer, after the harvest and in the late autumn.

There are two common methods of monitoring the population of varroa in hives that I use. They are:

- drone brood sampling;
- natural drop onto a sticky board.

Drone brood sampling is cheap and easy and is effected using a varroa fork. The fork is dipped into the brood surface and the drone pupae are impaled on the tines of the fork and the easily visible mites are noted. Once 100 cells have been uncapped the percentage of the cells affected is noted (not the total number of mites). If 5–10% of the cells are infected, then colony collapse may occur by the end of the season if not treated.

Varroa mite on pupae.

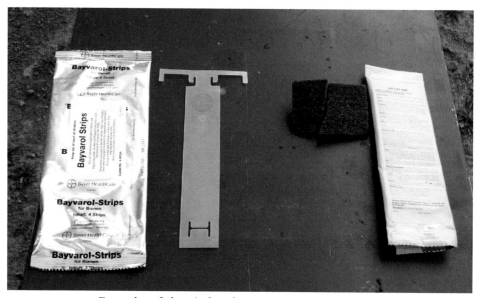

Examples of chemical and organic varroa treatments.

Natural mite drop sampling involves the use of a sticky board placed on the floor of the hive with a mesh screen placed on it to keep the bees from getting stuck. Proper mite drop floors can now be purchased which have a drawer in them which allow the beekeeper to carry out sampling without disturbing the bees. The sticky board is left for a set period of time or for an extended time depending on whether you want a quick assessment or a more accurate assessment. Three days is a good period for a quick assessment and I have found that it gives me a good idea of what is going on as the average daily drop is related to the overall mite population. Predicted varroa levels are not looking good if I find a daily average of more than one mite in the first inspection; 16

mites in July and in September 20 mites. Research shows that at these levels, colony collapse may occur before the end of the season.

Managing varoa

Managing varroa is all about limiting the number of mites in the colony to a level that won't cause harm and to this end a variety of 'weapons' are available. Chemical treatments are beginning to fail because the mite is becoming resistant to them and so organic treatments using oils, acids, fungi, bee breeding and so on are being encouraged. Because their life cycle accords more closely with that of drone development, varroa are attracted to drone brood and many beekeepers attempt a measure of control by removing this brood together with the mites. Be careful here. Altering hive dynamics is never a good idea but if you added *extra* drone comb to your hive or encourage the bees to draw out extra drone comb, then that would help but never underestimate the necessity for drones during the active season. Simply place a couple of shallow frames of foundation in the brood box. The space below the bottom bar of the frames will usually be filled up with drawn drone comb. This can then be sacrificed when full.

It is worth repeating that varroa control and management should always be coordinated with other beekeepers in your area and that means through the BKA which will have kept up to date with the latest thinking on this very complex subject.

Tropilaelaps clarae

Tropilaelaps clarae is another exotic mite that may reach the UK in due course. It is another small mite but is elongated in shape rather than crab shaped like varroa and like varroa it can cause colony collapse if untreated and is a notifiable pest in the UK.

Unlike the varroa mite, tropilaelaps cannot feed on adult bees because its mouthparts are unable to pierce the body wall membrane of the bees and here is a definite weakness for the mite. The mites depend on the developing brood for food, and move from the adult bees to feed on the larvae as quickly as possible after emergence and within one to two days. Any broodless period will therefore prove fatal to the mites and this could be a powerful weapon in managed bees.

Tropilaelaps clarae (right) and Varroa destructor. Note the easily spotted differences.

Tracheal mite: Acarine

You will hear in conversation much mention of the so called 'Isle of Wight Disease' and you will definitely hear it linked to what is known as Acarine disease. The acarine mite (*Acarapis woodii*) is a small mite that inhabits the prothoracic spiracle or breathing tube of the bee. These spiracles are in effect air inlets and enable the bee to breathe. Air enters the spiracles and is diffused by a branching system of smaller and smaller tubes which reach to all parts of the bee. The entrances to these spiracles are 'guarded' by hairs to prevent the ingress of objects and mites, but when a bee first emerges from the cell these hairs are still very soft. It is at this point that the little mite which is probably sitting on another bee at the time moves onto the young adult and enters the system and remains there. A build up of mites in the spiracle is detrimental to the bees' health and can cause problems. There are no real signs of this disease in the field. Some texts talk of symptoms such as crossed wings (K wing) and bees crawling at the hive entrance but these are probably not caused by acarine. These symptoms seem more like virus diseases of one sort or another. Whether the mite is connected to this by acting as a vector is another matter but certainly in the USA, it is a problem and is countered by using similar miticides to those used for varroa infestations. Treatment in Europe is generally unnecessary.

Ants and wasps

These are all members of the hymenoptera just like bees and all closely related, but this doesn't stop them raiding colonies whenever they get a chance. Wasps are a real nuisance after midsummer especially and can have dire effects on a colony. The only responses are to find the nest and destroy it and/or reduce the entrance to the hive so that the bees are better able to defend it. Also have a look at the colony because they might be weakened for some reason. Loads of wasps, or trails of ants entering a colony means that something is wrong. Check it out.

Hornets

Is it inevitable that the giant Asian hornet will reach the UK? They are in Brittany! *Vespa velutina* is capable of wiping out honey bee nests simply by simply sitting on the alighting board and decapitating any bee that attacks it. Once it has finished with the adult bees it will then remove the brood to feed its own young. Eastern honey bees have developed a way of dealing with it by balling it and raising the temperature in the ball to a degree more than the hornet can stand. It is a worrying thought and of course any sightings should be reported. The Food and Environmental Research Agency (FERA) has very good advice on this beautiful creature and gives advice on how to trap foundress queens. Take a look on the National Bee Unit website.

Small hive beetles (SHB)

You will hear a lot about these but at present they are not in Northern Europe. They hail from Africa where they are a minor pest of bees but they can have a devastating effect on bees in other countries. They literally destroy the colony. Both the larvae and the adults eat comb and brood and turn the whole lot into a slimy, decomposing, fermenting mess. They are amazingly adapted to life in with bees.

Small hive beetles with bees. Note the relative sizes.

If you open up a hive for an inspection, they can smell the hive odour from ten miles away and home in on it. They will fly with swarms and so when you capture a new swarm to increase your numbers you bring a veritable beast into your apiary. The adult beetles are about the third the size of the bees, black and look like standard beetles but note their club-shaped antennae and the six proto legs on the larvae unlike the legless wax moth larvae. If you see either of these beasts, notify your regional bee inspector straight away.

Other pests

Other pests such as birds do not really affect bees very much in northern Europe or the UK. Some will say that woodpeckers are a pest in the UK when bad winters strike, but even though their effect on an individual apiary may be considerable, their effect on bees overall is negligible. If you find woodpecker damage or are worried about this, many beekeepers use chicken wire to place over the hives. Woodpeckers can be persistent once they have found that they can get into a hive and the wire mesh must be far enough from the hive to prevent the bird getting near enough to the hive to drill the walls. A beekeeper friend of mine in the UK carries small pieces of wood in his 'kit' to cover holes made by woodpeckers using contact adhesive.

Bee eaters could reach the UK in the future with the effects of warming and have already been seen in the UK on rare occasions. I had them in Spain and they were interesting, pretty and quite deadly but again overall they didn't kill off colonies or anywhere near. One interesting observation was that when bee eaters were around my bees would leave the hive at low level and zig zag towards their destination almost like wartime merchant vessels trying to avoid submarines.

Conclusion

This chapter has attempted to give you a basic grounding in the problems that can afflict colonies and those that could in the future. They are important and they are serious. It is not a definitive lesson on the subject because that would be outside the scope of this book, but I hope that it will spur you on to reading and learning more about bee diseases and pests because they are just as much a part of it all as the bees themselves.

Remember that both your local beekeeping association and the regional bee disease inspector (employed by the national Bee Unit and part of DEFRA) will be only too pleased to support you in your fight against diseases. Please use them. Also remember that your local association will always be on hand to advise you about diseases, treatments and the law regarding them. All you have to do is keep up with developments.

Appendices: Useful Information for Beekeepers

Appendix 1: A beekeeping calendar

Remember that bees work according to the weather and the flowers, not the months so everything below can shift, but overall it gives a good picture of what you and your bees are meant to be doing during the year.

This isn't intended to be a comprehensive guide to nectar and pollen sources but I have mentioned the commoner flowering plants that may be of use to beekeepers and their bees each month. Make a note of all the plants in your area each month and build up your own calendar knowledge because flowering times depend on so many factors. Learn which ones give pollen and which give nectar – and/or both. I have just placed the flowers below in the months when they start their flowering period. Many of them can flower for several months, some for a shorter time. For a more comprehensive list, please read the *Beekeepers Field Guide*.

When you read the lists of flowers for each month, you can see just why the bees operate their calendar as they do, slowly starting and then a mad spring explosion in numbers before the decline. Just like the plants.

January

The bees: The bees are most probably in the midst of their winter cluster depending on the temperature outside the hive. There is little activity except on a warm day (about 45–50°F) when the workers will take cleansing flights. There are few if any drones in the hive (although I have seen drones in midwinter in the cluster). The queen may start laying.

The beekeeper: Keep an eye on your hives to ensure that snow isn't blocking the entrance or that flooding isn't affecting the hives. Other than that, little work is required from you at the hives. Winter is the time for reading, meeting other beekeepers in the association and winter projects such as building the solar wax extractor. Order new queens or nucs if required, from a reputable supplier.

Plants: Almond in warmer areas.

February

The bees: The queen will begin to lay more eggs each day. Workers will take cleansing flights on mild days.

The beekeeper: See January. Ensure that everything you are going to need for the new beekeeping year has been ordered or has been built by you. Remember the spring explosion! You will need everything ready. Do it now.

Plants: Almond, crocus, gorse, wild rose, broom, mahonia.

March

The bees: Right at the end of winter, the bees may have eaten their stores and there are still no new floral sources for them yet so really keep an eye on the hives to make sure the bees are flying well. This is the month when colonies can die of starvation. The queen will be increasing her rate of egg laying and more brood means more food consumed. In some areas early fruit blossom will be flowering and pollen sources such as gorse and willow will provide an early boost.

The beekeeper: Early in the month, on a nice, warm day and bees are flying, carry out a quick stores check. Have a look under the lid and see if there is any capped stores. If you do not see any sealed honey in the top frames, you may need to begin feeding and keep this going until the bees are bringing in their own. Remove mouse guards and place a queen excluder on top of the brood boxes.

Plants: Acacia, snowdrop, honeysuckles, aconites

April

The bees: The weather begins to improve and there will be more early blossoms for both nectar and pollen. You will see the bees begin to bring pollen into the hive. The queen is busily laying eggs, and the population is growing fast and you should start your first box reversals if necessary. Really keep an eye on this expansion. Drone rearing will be in full swing.

The beekeeper: Ensure that the colony is building up normally. If not, look for the problem and seek advice. You can now carry out your first full inspection. (See Chapter 5.) Keep a very wary eye open for signs of swarming (Chapter 6). Treat for varroa (Chapters 5 and 10).

Plants: Fruit trees, mustards, dandelions, crocus, berberis, currants, horsechestnuts, chickweed, henbit, leopard's bane.

May

The bees: This is probably when you started beekeeping and collected and installed your bees. If you bought a nuc or obtained a swarm, they will be young colonies and some of the information below won't apply. Read Chapter 5 for this but in your second year, use this calendar. Prime swarming season starts now. The bees should be storing honey at a fast rate and the queen will be laying equally fast. The hive will be really a mass of activity.

The beekeeper: You will be managing your small colony according to Chapter 5. But for a normal beekeeping year now is the time to add a queen excluder, and place honey supers on top of the top deep. Watch out for signs of swarming. Keep up your box reversal routine and if you really want to prevent swarming carry out an artificial swarm. Keep an eye out for disease at all times. Put out bait hives if required.

Plants: Dandelion, poppy, buddleia, ceanothus, firethorn, rosemary, rock roses, aparagus, blackberries, chives, sycamore, blackthorn, oilseed rape, raspberry, Lucerne, ajuga, holly, limes, cucumber.

June

The bees: If your colony hasn't swarmed make sure that the bees have room for stores and brood rearing. There can often be a nectar gap here before summer flowers come out so be careful to ensure your bees don't starve and watch out for robbing.

The beekeeper: Inspect the hive regularly. External inspections should be a constant feature of your routine. Internal inspections should be regular to make certain the hive is healthy and the queen is present. Add honey supers as needed. Don't give up on looking for signs of swarming. It is rare now but can still happen.

Plants: Borage, melons, pumpkin, sunflower, thyme, white clover, alyssum, basil, elder, lemon balm, oregano, mint, vipers bugloss, hebe, mock orange, laurel, catnip.

July

The bees: Summer flowers may be coming out now giving a welcome nectar source boost. If the weather is good, the nectar flow may continue throughout the month. Play flights will be an afternoon feature and on very warm evenings masses of bees may gather at and around the entrance.

The beekeeper: Continue your normal inspection routine to assure the health of your colony. Add more honey supers if needed and start thinking about the harvest.

Plants: thistles, cotoneasta, fireweed, germander, heather (*calluna vulgaris*), knapweed, mallow, marigold, Virginia creeper, clematis, daisy, potentilla.

August

The bees: It all starts to slow down a bit now but not yet too significantly. As usual, the pace of the hive depends on nectar flows. There is now little chance of swarming but there is a risk of robbing and wasps may become a serious nuisance now that they have changed their diet.

The beekeeper: Just keep an eye on the stores and keep a general eye on the colonies, looking out for problems, knowing their local flows. Some beekeepers will harvest towards the end this month. Honey shows start.

Plants: Eucalyptus.

September

The bees: The drones start to disappear. The rate of egg laying is reducing fast and so the population is dropping.

The beekeeper: Many beekeepers harvest their crop this month. If so, carry out all post-harvest activities required (Chapter 8). Remember to leave the colony with sufficient honey for winter (Chapter 9) and if not, then feed. Check for the queen's presence. If a hive is weak (but otherwise healthy), you can unite it to a stronger hive or another weak hive to make up numbers.

Plants: Goldenrod, ivys.

October

The bees: The colony is preparing for winter now. The queen's egg laying rate has been very much reduced and most if not all the drones have been expelled.

Robbing can be a problem now and the last of the wasps can be a problem for the colony.

The beekeeper: Watch out for robbing. Place mouse guard at entrance of hive and remove queen excluders. Complete any required feeding and carry out winter preparations for the hive ensuring that there are no cracks in the hives and that the lids are on securely. Place a brick if necessary on each lid.

Plants: Ivys

November

The bees: There will be little activity now. The bees may start to cluster depending on the weather.

The beekeeper: Store your equipment away for the winter and continue to carry out external inspections to make sure the hives are secure from the elements.

December

The bees: See November. The queen might just start laying again this month but don't open the hive. Just leave well alone for the time being.

The beekeeper: Not much to do now except your periodic external inspections and get on with your planning, reading and projects.

Appendix 2: What to do With it All

Using your well earned hive products

So hopefully you now have honey and maybe even some beeswax to show for your efforts over the year and you can congratulate yourself on being a fully fledged beekeeper. I'm sure you will agree that it was great fun and a huge learning experience. But now what? What can you do with them? This is really beyond the scope of this primer so my recommendation is that you read the specialist books on the subjects of cooking with honey, wine and mead making and also books on health products from honey and wax. It doesn't stop there either. There are many books on wax crafts for your pleasure and most of your products are easily saleable at fairs, farmer's markets and craft markets. Take a look at the reading list in Appendix 3 for ideas and also look at bee supply company catalogues. They open quite an exciting world that is entered into by many beekeepers – decorative candles, polishes and creams, beauty products and much more.

Now for a short taster

Because of the wealth of excellent books on these subjects, my intention at the end of this book is to give you an idea or a taster of what you could do with hive products and to encourage you to forage further on your own if you like the ideas presented. My favourite is cooking and in this I excel, but despite heroic efforts, I have never made decent mead and I advise you to purchase one of the books on the subject mentioned in the 'further reading' section. I have made dip candles and various small wax figures using rubber moulds and this is a fun thing to do especially if you have kids. Moulds and various pigments can be purchased from the beekeeping supply stores. I once made a series of Madonna and child figures and brushed them with an ageing pigment. They looked brilliant and I sold them in Spain along with my honey. People literally queued up for them. Lets' look at some more ideas.

Culinary ideas

Again, just a very few ideas to get you started. Many beekeepers use honey in cooking and there are some wonderful recipes from cakes and breads through to tfaya – so let's look at a few of recipes that use honey and let's start with tfaya to show you how versatile honey is as a culinary ingredient.

Tfaya

This Middle Eastern dish is a really delicious mix of honeyed caramelised onions and raisins called tfaya which is served as a topping on a meat dish such as lamb. The meat which should be braised until really tender is placed on a bed of couscous and then topped by this mouth watering mixture. You won't be disappointed with this one. The mix of flavours is sublime and the ingredients are few.

INGREDIENTS
    2 lb onions
    1 cup water
    2 tbsp butter

1 tbsp extra virgin olive oil
2 tbsp honey
1 ½ tsp cinnamon
1 cup raisins, soaked in water for 20 minutes
1 cup almonds

METHOD

Cut the onions in half and slice them. Put them in a pan with about 1 cup water. Put the lid on and cook, covered, over low heat (they will steam) for about ½ hour, until the onions are very soft. Remove the lid and cook until the liquid has evaporated. Add the butter and oil and cook until the onions are golden. Stir in the honey and cinnamon, the drained raisins, and a pinch of salt and cook 10 minutes more, or until the onions caramelise and become brown.

Toast the almonds in a dry frying pan or fry them in a drop of oil until golden, turning them over. Coarsely chop about half of them and sprinkle over the meal. I eat it with couscous.

Fesenjan-e Bademjan – Eggplant (Aubergines) with Pomegranate Braise

A dish from Persia (or is it Iran?)

This recipe may need a couple of specialist ingredients but if you can get hold of them it is worth the effort. If done well it will impress your dinner guests especially if they are vegetarian. I had trouble keeping the egg plants from being too wet after soaking them. In the end I decided not to soak them and they were still fine. In this recipe, 1 cup = 8 oz or 12 tablespoons.

INGREDIENTS

5 eggplants (2lb) or 2 large regular eggplants
6 tbsp olive oil
1 medium size onion, peeled and sliced
1 tbsp honey as needed
2 cloves garlic, peeled and chopped
2 tsp cumin
¼ tsp cinnamon
2 tsp salt

1 tsp freshly ground pepper

½ tsp turmeric

1 tsp crushed red pepper

1 cup chopped fresh parsley and cilantro (fresh oregano if you can't get cilantro)

½ cup chopped fresh mint

2 cup (½ lb) toasted walnuts

½ cup pomegranate paste diluted with 2 ½ cups water or 3 cups pomegranate juice

GARNISH

1 cup cilantro leaves, basil or parsley

1 cup pomegranate seeds (about 2 pomegranates)

fried eggplant rounds, fried cilantro or basil leaves and fried onion rings

METHOD

Peel the eggplants and remove the stems. Salt, let sit for at least 20 minutes, rinse and squeeze out any water. Slice into 1 inch thick rounds.

Heat 4 tbsp oil in deep frying pan over medium heat. Add the eggplant, and saute on all sides for about 15 minutes or until golden brown. Remove the eggplants from the skillet and drain on paper towels and set aside. Add the remaining oil to the pan and reheat over a medium heat.

Add onion and sauté until golden brown (10–14 minutes). Add garlic for the last few minutes of browning the onions and continue cooking. Add cumin, cinnamon, salt, pepper, turmeric, crushed red pepper, parsley, cilantro (or lemon basil or fresh oregano) and fresh mint and sauté for another 5 minutes. Remove from heat and set aside.

Grind walnuts in food processor until very fine and combine walnuts with diluted pomegranate paste and honey. Stir until sauce is smooth.

Pour sauce into the pan. Return eggplants and seasoning to skillet, reduce heat to low, cover and simmer mixture for 30 minutes or until eggplants are tender. If the sauce is too sour, add more honey to taste.

Spoon onto serving dish or plates, and garnish with herb leaves and pomegranate seeds, or any combination of the optional garnishes of fried eggplant rounds or crispy onions.

Serve with couscous or rice and a tomato and onion salad using as a dressing olive oil and vinagre de Jerez (Sherry vinegar). My family has couscous but I just like a stick of bread.

Green beans with garlic and honey

This is a very simple dish and tastes excellent if you use a strong-flavoured honey and was shown to me by a friend in New Zealand.

INGREDIENTS

> French beans preferably freshly picked
> 1 tbsp olive oil
> 2 cloves garlic finely chopped – or more if you like garlic
> 2 tsp strong honey. If you can get hold of some manuka honey then that's ideal.

METHOD

Saute the beans in the oil over a high heat for a few minutes. Add the garlic and toss for a bit longer to get the garlic very slightly browned and then add the honey and toss for another 30 seconds. Delicious and can be served with either of the above dishes.

Egg plant and honey

You will need a couple of small aubergines, some olive oil to fry them in and some honey and that's all but it's the way you cook them that counts.

METHOD

Slice the aubergines thinly. Place the slices on a paper towel and pat them dry and slightly salt them.

Heat up a frying pan to a very high heat. Those cast iron ones are best. Add some olive oil to the pan and heat to high. (Olive oil can reach a much higher heat than other oils without burning.)

Once dry, place the aubergine slices in the oil and fry on both sides very quickly until they brown but don't burn. Place them on a plate and drizzle with honey and that's it. Simple and delicious.

Finally, my overall favourite from my beekeeping days in Spain and used as a Christmas treat. Try it with a slug of Agua Diente or Anise.

Turron

INGREDIENTS

    100 g rosemary honey

    100 g caster sugar

    2 large egg whites beaten until stiff

    175 g blanched, lightly toasted almonds

    75 g roasted hazelnuts

    Rice paper

METHOD

Gently heat the honey and sugar until just boiling. Remove from heat and stir in the egg whites constantly beating the mix. Return to heat until it becomes toffee sticky and then stir in the nuts.

Line a baking tray with the rice paper and pour in the mix. Cover with more rice paper and place a weight on it for a few hours until it is well set and cool.

This really is delicious but try not to burn it as I was always doing.

Those recipes are just to show you how versatile honey can be in cooking and I urge you to experiment. Honey isn't all about honey cakes and baking, it can be used for many savoury dishes as well, giving that extra dimension of sweetness. But of course honey baking is a very popular pastime among beekeepers. Be careful when using honey for baking though. In baking, 1 cup or 12 oz of sugar should be replaced with only ¾ of the amount of honey and because it is more liquid, you need to reduce the amount of liquid used in the recipe.

    For more recipes using honey, go to one of the many websites thrown up by Google that have an abundance of honey recipes or follow the monthly recipes in APiSUK (www.apisuk.com). They are a bit more unusual.

Honey for health

There are many ways to produce face creams and scrubs using both honey and wax and most can be kept and re-used. Honey can have a dramatic effect on hair and can actually strengthen hair and make it glossy. For a simple hair tonic just mix a teaspoon of honey with a litre (2 pints) of warm water and rinse through your hair after shampooing. Do this daily and don't rinse out.

A really good tonic

For a first class general tonic, mix a tablespoon of honey with the same of a good cider vinegar (preferably organic). Drink this each morning. Honey on its own has very many medicinal properties and clinical trials have shown that buckwheat honey is better for coughs than many over-the-counter treatments. Manuka honey has such a large range of antibacterial and other properties that it is even used by the National Health Service for treatments for wounds and burns. Obviously, any treatments should be checked out with your doctor but I would suggest that honey won't do you any harm. In the UK you probably won't get any manuka or buckwheat honey (although there is a tiny manuka honey production from Cornwall which sells at £55 per jar) but I firmly believe that most honeys have medicinal properties of one sort or another.

Wax

Now what about wax? What can be done with this versatile hive product? Here are a few ideas apart from candle making and I start with the most obvious product and that's polish – the sort that smells like a real furniture polish should.

Furniture polishes

INGREDIENTS

   280 g (10 oz) clean beeswax (from your solar wax extractor)
   ½ litre (1 pint) of turpentine – Portuguese real stuff, not substitute.

METHOD

Shred the wax into a bain marie and stir until melted. Stir in the turps having slightly warmed it by standing the turps bottle in warm water.

Pour into small tins (which can be purchased from craft shops and bee supply companies) and allow to cool. If the polish is too soft for your purposes, next time reduce the amount of beeswax by 2 oz and add 2 oz of carnuba wax. This can also be purchased from bee supply companies and craft shops and is commonly added to beeswax for this purpose by beekeepers. It comes from a Brazilian tree and is harder than beeswax. This polish smells good and rich and can readily be sold. Labels for the polish tins can also be purchased and customised to your brand.

A softer polish or cream can be made by adding 60 ml (2 fl oz) liquid soap and 250 ml (9 fl oz) of warm water to 120 g (4.5 oz) of beeswax, 500 ml (1 pint) of turpentine and a small amount of pine oil (60 ml (2 oz). Again, this gives off a rich odour and is very effective. Remember to be very careful when heating wax and oils and always heat indirectly. Never put a flame anywhere near the business.

Wax can be used to make a variety of products including:

- decorative candles and ornaments
- lip balm
- cosmetics and medicinal creams
- foundation for new honeycomb in hives
- slippage prevention for belts in vacuum cleaners, sewing machines, etc.
- waterproofing shoes, fishing lines
- lubricant for doors, windows, tools
- wax for skis, toboggans, bow strings
- creating a freely moving surface on irons and frying pans
- furniture polish
- soap making.

So why not get started and build your own sales empire.

As I said earlier, these are just taster ideas to get you going. For many more craft recipes take a look at the book *Beeswax* by Ron Brown and the *Beeswax Craft* book both mentioned in the reading resources list. It's well worth looking into the world of wax and honey alternatives for cooking, health and beauty and other products and these specialist books will tell you all about it.

I hope this short section has given you some ideas of what you can do with your honey and I invite you to investigate further both by reading and through your beekeeping association. I'm sure you won't regret it.

Appendix 3: Taking it Further

Further general beekeeping books

A Practical Manual of Beekeeping: David Cramp. Spring Hill. Latest reprint 2010. This book is a more comprehensive and advanced beekeeping manual than this current book. It will take you into the realms of genetics, queen breeding and explores beekeeping around the world.

The Beekeeper's Field Guide: David Cramp. Spring Hill, 2011. This is not a 'how to keep bees' book but instead, an operational beekeeping manual that will tell you what to do in any given circumstance when you are out in the apiary. For example, it gives you a point by point hive trouble shooting guide and a point by point queen trouble shooting guide. It gives lists of flowers, trees and crops useful for bees and advises on whether they are nectar or pollen plants for bees and how much nectar they give. It features a very detailed disease field identification guide and advises on exactly what to do about each disease. Throughout, it advises you of the advantages and disadvantages of each manipulation you carry out. Generally speaking it is the only guide of its kind and will prove very useful to you.

The Hive and the Honey Bee: Ed. Joe Graham. Dadant & Sons, Dover Publications Inc. This is an ideal Christmas present. It is a large book – 465 pages and covers just about everything in beekeeping by a variety of well-known authors. It was first published over 150 years ago in the USA by Dadant himself and ever since then it has been one of the classic manuals. It is heavily slanted towards US beekeeping and methods but is useful for beekeepers all over the world.

At the Hive Entrance: H. Storch. European Apicultural Editions, 1985. This book will give you a very good understanding of what is going on in the hive using the external inspection method. It is a very good book and gives some excellent advice but remember, it is very dated and Storch did not have to deal with varroa, CCD and other exotic diseases. Use the book wisely and ask advice from your association about your observations. Don't get caught out.

For Gardeners

The Bee Garden: How to Create or Adapt a Garden To Attract and Nurture Bees: Maureen Little. Spring Hill, 2011. An excellent, modern book which shows you exactly how to plant for bees and of course other beneficial, pollinating insects. By avoiding the big, showy multi-petalled plants that abound in nurseries these days you can make a real difference to your garden and remember that there are about 3 million hectares of gardens in the UK – a larger area than all the national parks put together, so they are extremely important for bees.

Plants and Beekeeping: F.N. Howes. Faber and Faber, 2nd revised edition 1979. This is the classic book by an expert on the subject. My copy is the 1945 edition and it is a bit dated but only in that many new varieties are not included. My copy is interesting in that it is inscribed (in pencil): To Captain Bolster from June and James. Oct '46. I have always wondered who they were!

Other hive types/beekeeping methods

The Barefoot Beekeeper: P.J. Chandler. Covers most aspects of using top bar hives, both Warre and Kenya types and the book goes through all the manipulations necessary to successfully keep bees in these hives. Also available as an E-book: www.biobees.com.

The Bee-friendly Beekeeper: A Sustainable Approach: David Heaf. An insightful exploration of modern beekeeping practices, and how they can be improved for a more sustainable and bee-friendly approach.

Meditation and the Art of Beekeeping: Mark Magill. Beekeeping is not just fun, educational and sustainable – it may also be the path to enlightenment. A gentle book. Very good and very Buddhist. I loved it.

Rearing your own queens

Queen Rearing Simplified: Vince Cook. Northern Bee Books, 2008. This book which was re-issued in 2008 by NBB was the one I turned to when I decided to rear my own queens. At the time, I was unable to afford one of the queen rearing kits now available and bought this book instead. It makes the whole deal easy and simple and explains the process so well that anyone can do it. If you decide that instead of buying queens to re-queen your hives you would like to have a go yourself, this is the book for you. Since those early days, I have never seen another book that makes the process so easy for new beekeepers.

Queen Bee: Biology, rearing and breeding: David Woodward, head of the apiculture department at Telford Rural Polytechnic, New Zealand. Northern Bee Books, 2010. This book takes you a lot further into queen biology, rearing and breeding for those who want to really get into the subject. The book gives a lot of advice on a wide range of beekeeping activities.

Pollen

Honey Identification: Rex Sawyer. Cardiff Academic Press, 1998. This book takes you into the identification of honey by pollen analysis. It tells you how to do it and what equipment you will need to become a honey detective. By using this type of analysis, cheap honey from China claiming to be 'Best Dorset' honey can be unmasked. It is an interesting subject and the more you get into it, the more you learn about the honey bee's intimate relationship with the local flora.

A Colour Guide to the Pollen Loads of the Honey Bee: W. Kirk. International Bee Research Association, 2006. This book is of interest to those beekeepers who want to know which plants their bees are foraging on for pollen. It helps you build up a picture of the plants in your area and makes you far more aware of floral sources and the bee plant interactions. A very good book.

Bee biology and microscopy

Form and Function in One Honey Bee: IBRA, 2003. This is probably the best book on the subject. Lavishly illustrated with over 300 colour illustrations, photographs and diagrams, this book is an up-to-date guide to the biology of the honeybee. It is an introduction for students, beekeepers and others and for those interested in this aspect of beekeeping it is essential.

Anatomy and Dissection of the Honeybee: Harry Dade. IBRA, 1994, revised in 2009. This is still the classic on the subject. If you are interested in microscopy why not research the honey bee. Dade tells all in a very clear and concise manner and it was my guide (and every other student's) when I was carrying out bee research at Cardiff.

Philately

There are many philatelists around the world who collect stamps with beekeeping related images and themes. See: http://www.lungau.de/bienenmar-ken/frames/eng_frame.htm. Take a look at this site which is German but just click on the Union Jack to put it into English.

Beekeeping history and bees and mankind

Bees and Mankind: John B, Free. George Allen and Unwin, 1982. This book gives a global look at bees and how they have interacted with mankind throughout history. Professor Free who gave excellent lectures during my year at Cardiff was an international expert on bees.

The Hive: Bee Wilson. John Murray, 2005. The story of the honey bee and us. The appropriately named Bee Wilson explores our relationship with the bee over centuries and for anyone interested in beekeeping history and lore, this is for them.

The Sacred Bee in Ancient Times and Folklore: Hilda Ransome. Dover Publications Inc, 2004. This book goes even further back in time to the ancient Egyptians and the Hittites as well as developing customs and practices in the UK and Europe. It is comprehensive, unusual and really interesting for those wanting beekeeping history and myth.

Producing and using other hive products

Producing Royal Jelly: A Guide For The Commercial and Hobbyist Beekeeper: Dr Ron van Toor. Bassdrum Books Ltd, 2006. This is the most up-to-date and complete guide to producing royal jelly currently available. It doesn't matter whether you have one hive or one thousand or more, Ron's clear text, illustrative diagrams and colour photos will take you step by step through every stage of producing this healthy and profitable hive product.

Beeswax: Ron Brown. Bee Books New and Old, 1989. This is a classic on beeswax and should appeal to all beekeepers. It gives you the history of wax and gets as scientific as you like yet gives you many ideas about how to render wax using easily constructed wax extractors as well as how to show your wax, make candles and gives some excellent ideas for creams and even how to make turner's cement! This is a must read book for anyone interested in this aspect of beekeeping. I have the second edition (1989) but there is a third (revised) edition which came out in 1995.

Honey Wines and Beers: Clara Furness. Northern Bee Books, 1987. Written by an expert in her field, this book tells you exactly how to brew some beautiful drinks.

Beeswax Crafts, Candlemaking, Modelling, Beauty Creams, Soaps and Polishes, Encaustic Art, Wax Crayo: Search Press, 1997. Another excellent book that will give you loads of ideas when using wax in some surprising ways.

The Book of Honey: Margaret Briggs. Hermes House, Anness Publishing, 2010. A very useful book for anyone wanting clear, simple recipes for all manner of beauty treatments using honey, cookery recipes, healing recipes and so on. Well worth getting hold of a copy.

Honey. BeesOnLine Recipes. Get the book by Maureen Maxwell, Tandem Press, 2003 or go online at: www.beesonline.com and go to the recipe section. The recipes here are really to my taste and for me this is probably the best honey recipe book around. Their honey roasted tomatoes or green beans with honey and garlic are superb.

Apitherapy

The Medical Aspects of Beekeeping: Harry Riches. Northern Bee Books, 2009. A re-issue of a very interesting book and well worth reading for beekeepers of all experience levels. This is a good introduction to the increasingly important world of apitherapy. (See also IBRA below.) Remember, apitherapy is moving rapidly into mainstream medicine now.

Follow this up with an online course from some of the most expert apitherapists in the world. See: http://www.mdbee.com/apitherapy/aic.html. A full course is offered for 200 Euros which is exceptional value and is backed by Apimondia, the beekeeping world's governing body.

Magazines

There are two general beekeeping magazines available in the UK aimed at UK beekeepers and both are worth investing in as their editorial does have differences in focus and as far as I can see they complement each other rather than competing. There are some overlaps but nothing significant.

BeeCraft: An all colour UK monthly magazine of interest to beekeepers of all experience levels. This magazine will keep you up to date on all aspects of beekeeping in the UK and will keep you in touch with what is going on in the British Beekeepers Association. See: http://www.bee-craft.com/

The Beekeeper's Quarterly: Published by Northern Bee Books. Another all colour magazine that covers all aspects of beekeeping and includes articles from correspondents all over the world. Excellent value for money. See: www.beedata.com/bbq.htm.

Two more UK-based magazines offer more specialised views of beekeeping and show just how vast this subject is and just how far it can take you.

Bees for Development Journal: Another UK magazine published by the charity, Bees for Development based in Monmouth. It is aimed at beekeepers in developing countries and so tends to specialise in low-cost solutions to beekeeping problems in these countries. It shines an interesting light on a very different aspect of the craft to that of UK beekeeping and you can learn a lot from this. Also, you can buy a subscription and have your copy sent to a needy beekeeper overseas who otherwise wouldn't be able to afford a copy. See: www.beesfordevelopment.org

Bee World: Published by the Cardiff based International Bee Research Association (IBRA), this journal offers a global perspective to beekeeping as well as keeping tabs on more local issues. Members of the association receive the journal as part of their membership. IBRA also publishes the *Journal of Apicultural Research* and the *Journal of ApiProduct and ApiMedical Science*. These two very specialised journals are peer reviewed and if you enjoy the science of beekeeping and/or the enormous subject of Apitherapy then they are for you.

Beekeeping courses

Most, if not all of the beekeeping associations will offer courses at various levels but mostly at beginner level and you should contact your local association for details. Other courses will cover such subjects as queen rearing, microscopy and other more specialist topics as and when they get in a visiting expert. The BBKA itself offers correspondence courses at various levels. These courses cover the Basic Certificate as well as the seven modules that form the route to the Intermediate Theory Certificate and then on to the Senior Theory Certificate. There is also a course which covers the learning for the oral aspect of the Microscopy assessment.

The National Diploma in Beekeeping (NDB): The following information is taken from the website: http://national-diploma-bees.org.uk. The National Diploma in Beekeeping exists to meet a need for a beekeeping qualification above the level of the Certificates awarded by the United Kingdom National Beekeeping Associations. The Diploma is the highest beekeeping qualification recognised in the United Kingdom and its holders are generally well known figures within beekeeping education. Why not work towards this high qualification?

Some very useful organisations and websites

British Beekeepers' Association: www.bka.org.uk

Scottish BKA: www.scottishbeekeepers.org.uk

Welsh BKA: www.wbka.com

Ulster BKA: ubka.org

The Federation of Irish Beekeepers' Associations: www.irishbeekeeping.ie

The National Bee Unit: Provides 'BeeBase' which is the National Bee Unit website offering a wide range of free beekeeping information for UK Beekeepers. Effectively your national beekeeping authority. See: www.nationalbeeunit.com.

IBRA: The Cardiff-based International Bee Research Association. Based in Cardiff it promotes the scientific study and understanding of bees. See: www.ibra.org.uk.

Apimondia is the International Federation of Beekeepers' Associations and other organisations working within the apiculture sector. Apimondia exists to promote scientific, technical, ecological, social and economic apicultural development in all countries. It organises the two yearly international bee shows held in different countries which are well worth a visit. See: www.apimondia.com.

Dave Cushman's website: Just about everything you wanted to know about beekeeping is here. Dave recently died but his excellent website is being kept up. www.dave-cushman.net

The Virtual Beekeeping Gallery: www.beekeeping.com opens up the world of beekeeping to you. Anything you need to know around the world, this is the place.

Bee Science News: Keeps you up to date with bee research and findings in an easy to read format. Learn about new discoveries. www.apisuk.com

Appendix 4: Glossary of Terms

So that you are not stumped when new terms are introduced at meetings or in conversation with other beekeepers, here are the words you need to know:

Alighting board – a small projection or platform at the entrance of the hive.

American Foul Brood (AFB) – a brood disease of honey bees caused by the spore-forming bacterium, *Bacillus larvae.*

Anaphylactic shock – constriction of the muscles surrounding the bronchial tubes of a human, caused by hypersensitivity to venom and resulting in sudden death unless immediate medical attention is received.

Apiary – colonies, hives, and other equipment assembled in one location for beekeeping operations; bee yard.

Apiculture – the science and art of raising honey bees.

Apis mellifera – scientific name of the honey bee.

Bait hive – a hive placed to attract stray swarms.

Bee blower – an engine with attached blower used to dislodge bees from combs in a honey super by creating a high-velocity, high-volume wind.

Bee brush – a brush or whisk broom used to remove bees from combs.

Bee escape – a device used to remove bees from honey supers and buildings by permitting bees to pass one way but preventing their return.

Beehive – a box or receptacle with movable frames, used for housing a colony of bees.

Bee space – ¼- to ⅜-inch space between combs and hive parts in which bees build no comb or deposit only a small amount of propolis.

Beeswax – a complex mixture of organic compounds secreted by special glands on the last four visible segments on the ventral side of the worker bee's abdomen and used for building comb.

Bee venom – the poison secreted by special glands attached to the stinger of the bee.

Bottom board – the floor of a beehive.

Brace comb – a bit of comb built between two combs to fasten them together, between a comb and adjacent wood, or between two wooden parts such as top bars.

Brood – bees not yet emerged from their cells: eggs, larvae, and pupae.

Brood chamber – the part of the hive in which the brood is reared; may include one or more hive bodies and the combs within.

Capped (or sealed) brood – pupae whose cells have been sealed with a porous wax cover.

Cappings – the thin wax covering of cells full of honey; the cell coverings after they are sliced from the surface of a honey-filled comb.

Castes – the three types of bees that comprise the adult population of a honey bee colony: workers, drones, and queen.

Cell – the hexagonal compartment of a honey comb.

Chilled brood – immature bees that have died from exposure to cold; commonly caused by mismanagement.

Cluster – a large group of bees hanging together, one upon another.

Colony – the aggregate of worker bees, drones, queen, and developing brood living together as a family unit in a hive or other dwelling.

Comb – a mass of six-sided cells made by honey bees in which brood is reared and honey and pollen are stored; composed of two layers united at their bases.

Comb foundation – a commercially made structure consisting of thin sheets of beeswax with the cell bases of worker cells embossed on both sides

Creamed honey – honey which has been allowed to crystallise, usually under controlled conditions, to produce a tiny crystal.

Crystallisation – see 'Granulation.'

Dequeen – to remove a queen from a colony.

Dividing – separating a colony to form two or more units.

Drawn combs – combs with cells built out by honey bees from a sheet of foundation.

Drifting of bees – the failure of bees to return to their own hive in an apiary containing many colonies.

Drone – the male honey bee.

Drone comb – comb used for drone rearing and honey storage.

Drone layer – an infertile or unmated laying queen.

Dwindling – the rapid dying off of old bees in the spring; sometimes called spring dwindling or disappearing disease.

Dysentery – an abnormal condition of adult bees characterised by severe diarrhea and usually caused by starvation, low-quality food.

European Foul Brood (EFB) – an infectious brood disease of honey bees caused by streptococcus p/u ton.

Extracted honey – honey removed from the comb by centrifugal force.

Fermentation – a chemical breakdown of honey, caused by sugar-tolerant yeast and associated with honey having a high moisture content.

Fertile queen – a queen, inseminated instrumentally or mated with a drone, which can lay fertilised eggs.

Field bees – worker bees at least three weeks old that work in the field to collect nectar, pollen, water, and propolis.

Frame – four pieces of wood designed to hold honey comb, consisting of a top bar, a bottom bar, and two end bars.

Fructose – the predominant simple sugar found in honey; also known as levulose.

Fumidil-B – the trade name for Fumagillin, an antibiotic used in the prevention and suppression of nosema disease.

Fume board – a rectangular frame, the size of a super, covered with an absorbent material such as burlap, on which is placed a chemical repellent to drive the bees out of supers for honey removal.

Grafting – removing a worker larva from its cell and placing it in an artificial queen cup in order to have it reared into a queen.

Grafting tool – a needle or probe used for transferring larvae in grafting of queen cells.

Granulation – the formation of sugar (dextrose) crystals in honey.

Hive – a man-made home for bees.

Hive body – a wooden box which encloses the frames.

Hive stand – a structure that supports the hive.

Hive tool – a metal device used to open hives, pry frames apart, and scrape wax and propolis from the hive parts.

Honey – a sweet viscid material produced by bees from the nectar of flowers.

Honeydew – a sweet liquid excreted by aphids, leaf hoppers, and some scale insects that is collected by bees.

Honey extractor – a machine which removes honey from the cells of comb by centrifugal force.

Honey flow – a time when nectar is plentiful and bees produce and store surplus honey.

Honey stomach – an organ in the abdomen of the honey bee used for carrying nectar, honey, or water.

Increase – to add to the number of colonies, usually by dividing those on hand.

Inner cover – a lightweight cover used under a standard telescoping cover on a beehive.

Invertase – an enzyme produced by the honey bee which helps to transform sucrose to dextrose and levulose.

Larva (plural, larvae) – the second stage of bee metamorphosis; a white, legless, grublike insect.

Laying worker – a worker which lays infertile eggs, producing only drones, usually in colonies that are hopelessly queenless.

Levulose – see 'Fructose'.

Mating flight – the flight taken by a virgin queen while she mates in the air with several drones.

Mead – honey wine.

Mellisococcus pluton – bacterium that causes European Foul Brood

Nectar – a sweet liquid secreted by the nectaries of plants; the raw product of honey.

Nectar guide – colour marks on flowers believed to direct insects to nectar sources.

Nectaries – the organs of plants which secrete nectar, located within the flower (floral nectaries) or on other portions of the plant (extrafloral nectaries).

Nosema – a disease of the adult honey bee caused by the protozoan Nosema apis.

Nucleus (plural, nuclei) – a small hive of bees, usually covering from two to five frames of comb and used primarily for starting new colonies, rearing or storing queens; also called a 'nuc'.

Nurse bees – young bees, three to ten days old, which feed and take care of developing brood.

Out-apiary – an apiary situated away from the home of the beekeeper.

Oxytetracycline – an antibiotic used to prevent American and European Foul Brood.

Package bees – a quantity of adult bees (2 to 5 lb), with or without a queen, contained in a screened shipping cage.

Paenibacillus larvae – the bacterium that causes American Foul Brood

Paralysis – a virus disease of adult bees which affects their ability to use legs or wings normally.

Parthenogenesis – the development of drone bees from unfertilised eggs.

Play flight – short flight taken in front of or near the hive to acquaint young bees with their immediate surroundings.

Pollen – the male reproductive cell bodies produced by anthers of flowers, collected and used by honey bees as their source of protein.

Pollen basket – Curved spines or hairs, located on the outer surface of the bee's hind legs and adapted for carrying pollen and propolis.

Pollen trap – a device for removing pollen loads from the pollen baskets of incoming bees.

Prime swarm – the first swarm to leave the parent colony, usually with the old queen.

Propolis – sap or resinous materials collected from trees or plants by bees and used to strengthen the comb, close-up cracks and keep the hive healthy.

Pupa – the third stage in the development of the honey bee, during which the organs of the larva are replaced by those that will be used by an adult.

Queen – a fully developed female bee, larger and longer than a worker bee.

Queen cage – a small cage in which a queen and three or four worker bees may be confined for shipping and/ or introduction into a colony.

Queen cage candy – candy made by kneading powdered sugar with invert sugar syrup until it forms a stiff dough; used as food in queen cages.

Queen cell – a special elongated cell, resembling a peanut shell, in which the queen is reared.

Queen clipping – removing a portion of one or both front wings of a queen to prevent her from flying.

Queen excluder – metal or plastic device with spaces that permit the passage of workers but restrict the movement of drones and queens to a specific part of the hive.

Queen substance – pheromone material secreted from glands in the queen bee and transmitted throughout the colony by workers to alert other workers of the queen's presence.

Rendering wax – the process of melting combs and cappings and removing refuse from the wax.

Robbing – stealing of nectar, or honey, by bees from other colonies.

Royal jelly – a glandular secretion of young bees, used to feed young brood.

Sac brood – a brood disease of honey bees caused by a virus.

Scout bees – worker bees searching for a new source of pollen, nectar, propolis, water, or a new home for a swarm of bees.

Self-spacing frames – frames constructed so that they are a bee space apart when pushed together in a hive body.

Slumgum – the refuse from melted comb and cappings after the wax has been rendered or removed.

Smoker – a device in which burlap, wood shavings, or other materials are slowly burned to produce smoke which is used to subdue bees.

Solar wax extractor – a glass-covered insulated box used to melt wax from combs and cappings by the heat of the sun.

Spermatheca – a special organ of the queen in which the sperm of the drone is stored.

Sting – the modified ovipositor of a worker honey bee.

Sucrose – principal sugar found in nectar.

Super – any hive body used for the storage of surplus honey.

Surplus honey – honey removed from the hive which exceeds that needed by bees for their own use.

Swarm – the aggregate of worker bees, drones, and usually the old queen that leaves the parent colony to establish a new colony.

Swarming – the natural method of propagation of the honey bee colony.

Swarm cell – queen cells usually found on the bottom of the combs before swarrning.

Uncapping knife – a knife used to shave or remove the cappings from combs of sealed honey prior to extraction; usually heated by steam or electricity.

Uniting – combining two or more colonies to form a larger colony.

Venom allergy – a condition in which a person, when stung, may experience a variety of symptoms ranging from a mild rash or itchiness to anaphylactic shock.

Virgin queen – an unmated queen.

Wax glands – the eight glands that secrete bees wax; located in pairs on the last four visible ventral abdominal segments.

Wax moth – larvae of the moths *Galleria mellonella* or *Achroia grisella*, which seriously damage brood and empty combs.

Winter cluster – the arrangement of adult bees within the hive during winter.

Worker bee – a female bee whose reproductive organs are undeveloped.

Worker comb – comb in which workers are reared and honey and pollen are stored.

Index

abdomen, 57–8, 63–4, 168

acarine, 128, 142

adrenaline injections, 21

adult bees, 132, 139, 141, 143, 167, 169,
    172
  diseases, 137

AFB (American Foul Brood), 41, 76, 95,
    125, 133–4, 135–6, 165

age, 7–8, 14, 22, 41, 47, 51

aggressive colonies, 100

allergic emergencies, 16

American Foul Brood *see* AFB

annual beekeeping association auction, 48

ants, 53, 96, 117, 143

anatomy, 57

apiary, 17, 44–6, 52, 69, 76, 83, 88, 98–9,
    102, 110, 118, 120, 144, 158, 165–6

Apimondia, 162, 164

Apis mellifera, 53, 67, 82, 165

apitherapy, 11, 162–3

artificial swarm, 88–9, 129, 148

assessment, quick, 141

associations, 45, 69, 74, 83, 105, 128, 139,
    157, 163, 164

At the Hive Entrance, 40, 158

attacking bees, 26

autumn, 76, 82–3, 105, 107, 109, 111, 113,
    115, 117, 119, 121, 130, 137

autumn honey, 42

bacteria, 11, 42, 76, 136

bait hives, 42, 93, 165

baking, 155

BBKA (British Beekeepers' Association),
    22, 29, 45, 80, 162, 163–4

bee
  number, 102, 127–8
  venom, 11, 16–17, 165
  vision, 53, 57
  year, 52, 60, 77

bee eaters,12, 144

beehives, 18–19, 26–7, 31–43, 69, 154, 165,
    186
  most common, 36
  types, 32–41

Bee Improvement and Bee Breeders'
    Association (BIBBA), 55

beekeepers
  amateur, 8
  bad, 29
  commercial, 14, 23, 97, 103, 117
  disabled, 23
  garden, 38
  local, 29, 32, 45, 47–8, 77, 106
  neighbouring, 41
  new, 4, 18, 29, 42–3, 44, 76, 159
  professional, 76
  prospective, 17, 23
  registered organic, 41

Beekeepers' Association *see* BKA

Beekeepers Field Guide, 41, 89, 102, 134,
    146, 158

beekeeper's operations manual, 41, 89

bees
  alfalfa leaf cutter, 10
  carnica, 55

173

forager, 51
gentle, 47, 66
gentle Greek, 55
honey, 10
Iberian, 55
Italian, 99
ligurian, 5
scout, 51 93, 101–2, 185
solitary, 10, 53
urban, 24
Bees for Development Journal, 162
beeswax, 34, 120, 127, 150, 156–7, 161, 165,
    166
Beeswax, 157, 161
Beeswax Crafts, 157, 161
biology, 53, 159–60
BKA (Beekeepers' Association), 17, 19, 22,
    48, 71, 79, 100, 115, 133, 141
board
   alighting, 96, 136, 138, 143, 165
   sticky, 139–40
books, 40, 157
bottom bars, 38–9, 83, 141, 167
boxes
   extra, 118, 125
   full, 109, 112
   nuc, 90–1, 127
   stacking, 38–9
breeding, 11, 53, 159
British Beekeepers' Association *see* BBKA
British National frame, 121
brood, 9, 19, 31, 48, 51, 70, 72–4, 86–8,
    110, 121, 123–4, 132, 138, 143–4,
    166
   box, 62
   chalk, 96, 136
   chilled, 136
   diseases, 133, 135–7
   food, 51, 77

frames, 70, 73–4, 87–8, 100, 102, 123,
    133, 135–6
   nest, 36, 70
bumblebees, 10, 53

candy, 44, 100, 124
   queen cage, 170
cappings, 108, 113–14, 166, 170–1
captured swarm, 62
career, 104, 139
   carpentry, 11, 35
castes, 49–51, 166
catching swarms, 61, 90–3, 127
CCD (colony collapse disorder), 137–9,
    158
cecropia, 54–5
cells, 9, 39, 49, 51, 64, 70–2, 74–6, 79, 84,
    101–2, 109, 112, 135, 139, 166–8
chemical-free beekeeping, 41
chromosomes, 82
classes, 44
clinical trials, 11, 156
clothing, 18–19, 21
cluster, 65, 123, 128, 146, 150, 166
colonies, 51, 124, 132–3, 166
   aggressive, 100
   characteristics, 48
   dead, 125
   equalizing, 86
   gentle, 48, 100
   healthy, 132–3, 135
   large, 50, 65, 73
   nasty, 100
   new, 64, 125, 169, 171
   parent, 170–1
   queenless, 101
   raiding, 143
   strong, 86–7, 102
   uniting, 89

weak, 86–7, 102
colony collapse disorder *see* CCD
colony swarms, 89
combs, 3, 34–5, 39, 92, 111–12, 181, 166
    brace, 34, 165
    natural, 42–3
    storage, 118
    worker, 172
commercial beekeepers, 35
competitions, annual beekeeping
    association, 13
container, honey storage, 107
cookery, 3, 13
cooking, 1, 150–1, 153, 155, 157
costs, 11, 18–20, 22, 122, 127
countryside, 23–4, 26, 29, 100
courses, 17, 163
craft shops, 156
creamed honey, 166
crocus, 128, 147–8
crops
    bees pollinate, 6
    good honey, 55
    produced excellent honey, 42
crown board, 78, 81, 122, 124
cup, artificial queen, 167

dance, 55–6, 92
dandelions, 57, 148
DCAs (Drone Congregation Areas), 52–3,
    84
dead bees, 67, 96, 138
    few, 89
    removing, 51
dead colonies, 125
DEFRA (Department for the Environment,
    Food and Rural Affairs), 17, 145
dequeen, 166
destroy ailing colonies, 134

dextrose, 168
diversity, genetic, 51–2
drone comb, 141
Drone Congregation Areas *see* DCAs
drones, 49, 52–3, 85, 96, 146, 149, 167
duties, nursing, 51
dysentery, 137

early spring, 67, 78, 128, 139
Eastern honey bees, 67, 82, 143
EFB (European Foul Brood), 41, 76, 133,
    136, 167
eggs
    fertilized, 82, 167
    laying, 84, 147
    rate of, 147, 149
    single, 49, 132
    unfertilized, 52, 82, 169
empty
    cells, 70, 74
    combs, 118, 172
entrance, 48, 61, 64–7, 81, 90–1, 95–101,
    103, 123–4, 137, 142–3, 146, 149, 165
epinephrine, 16
EpiPen, 16–18, 21
equalising colonies, 86
equipment, 18–21, 90, 106–9, 150
European Foul Brood *see* EFB
exchanging frames, 86
excluder, 74, 79
experience, 8–9, 30, 41–2, 45, 95, 99, 128,
    134, 171
expert, 5, 132, 158, 161, 163
external inspections, 41, 95, 120, 148, 150
extraction, 106, 114, 117, 171
extractor, 34, 105–8, 110, 112–14, 126
    electric, 112, 114
    solar wax, 126, 146, 156, 171

Family Apidae Honey, 54
fanning, 64, 97
farmers, 4–5, 102
feeders, 76, 90, 122
fences, high, 25
FERA (Food and Environment Research
    Agency), 8, 30, 143
fighting, 95, 98
flight
    cleansing, 146–7
    mating, 168
    play, 149, 169
flowers, 9–10, 24, 130, 146, 158, 169
flying, 64, 66, 97
forage, 24, 27, 56, 64, 97, 130, 151
foragers, 51, 56, 64–5, 87–8, 96, 102–3
foundation, 32, 34, 47, 62, 64, 88, 93, 108,
    127, 141, 157, 166
frame
    beehives, 18
    feeders, 76, 122
frames, 34, 38, 64–5, 102, 118, 167
    dripping, 105
    empty, 13
    exchanging, 86–7
    extra, 125
    extracted, 108, 117–18
    healthy, 133
    lifting, 20, 78–9, 81
French beekeeper Abbe Warre, 39
fructose, 167

gardeners, 2–3, 29, 158
gardens, 1, 4, 8, 24, 28, 50, 60, 98, 102–3,
    125, 158
    small, 24–5, 62, 118
genetics, 41, 100, 157
gentle queen, 100
Genus Apis Honeybees, 54

glands, 9, 64, 120
    nasonov, 63–4, 97
gloves, 21, 106, 135
gorse, 77, 147
governments, 5–6, 8, 30, 138
granulation, 166, 168

harvest, 105–15, 149
health, 6, 11, 28, 40, 142, 149, 155, 157
heat, 36, 153–5, 157, 171
history, 7, 12, 70, 160–1
hive
    entrance, 40–1, 65, 95, 96, 103, 123
    frames, 61
    inspections, 35, 41, 69, 78–80, 94
    lids, 68, 122–3
    problems, 94–104
    products, 4, 11, 130, 150–1, 161
    tool, 20, 45, 61, 75, 79, 106, 168
hives, 18–19, 31–43
    bait, 42, 93, 130, 148, 165
    cork, 42
    empty, 65, 91
    extra, 89, 125
    moving, 87, 102, 103
    multi-coloured, 10
    nucleus (nuc), 48, 66
    robbed, 98
    secondhand, 74
    siting, 24–7
    small, 48, 62, 169
    strong, 87, 118
    top bar type, 32, 38
    vertical stacking type, 39
    weak, 87, 149
hobbyist, 34, 35, 105
honey, 1–5, 13
    apple, 80
    boxes, 74, 111, 118, 125

buckwheat, 11, 156

capped, 111, 117

clear, 107–8, 113

comb, 39, 126

common UK, 114

early, 80

extracted, 105, 167

flows, 81, 107, 130, 168

heather, 114, 117

raw, 3

sealed, 147, 171

selling, 115

sterile, 108

storing, 34, 76, 115, 148

supers, 74, 78, 86, 109, 148, 171

surplus, 84, 87, 168, 171

tap, 106

testing, 117

using, 11, 13, 155, 161

Honey Identification, 159

Honey Wines and Beers, 161

honeycomb, 34, 36, 39

honeydew, 55, 168

honeysuckles, 147

hornet, 143

horses, 27

humblebees, 54

hygrometer, 117

Iberian honey bee, 55

IBRA (International Bee Research
   Association), 160

insects, 7, 10, 43, 44, 46, 49, 51, 58, 131

   pollinating, 10, 158

inspections, 35, 40, 68–70, 78–9, 81, 94–5,
   121, 129

insurance, 26, 29

International Federation of Beekeepers'
   Associations (Apimondia), 164

Irish Beekeepers' Associations, 164

Kenya top bar hive, 38

killer swarms, 44

kit

   small honey processing, 19

   swarm catching, 90

labels, 116–17, 156

Langstroth frame/hives, 23, 35, 39, 42, 121

Langstroth, Rev. L., 34

larva 54, 150, 183, 185

larvae, 49, 51, 70, 96, 132–3, 135–7, 142,
   159, 166, 168

laws, 28–9, 43, 145

levulose, 168

ligustica, 54

liquid honey, 4

live charges, 31

local association, 17, 19, 22, 31, 60, 77, 129,
   133, 145, 163

locations

   rural, 8, 26–7

   urban, 24–6

London Beekeepers Association, 24

London Stock Exchange (LSE), 24

magazines, 127, 162

manuka honey, 156

marking queens, 71

mating, 44, 49, 52, 96, 101, 168

mesh floors, 36–7, 90, 123

microscopy, 1, 10, 44, 57, 160, 163

mites, 67–8, 83–3, 129, 139–42

mouse guard, 123, 128, 147, 150

moving, 87, 102, 103

National Bee Unit (NBU), 17, 30, 143, 145,
   164

National Diploma in Beekeeping (NDB), 163

National hives, 39, 66

National size (British), 35, 36, 37

NBU *see* National Bee Unit

nectar, 5, 26, 49, 51, 58, 67, 70, 76–7

nectarines, 169

neighbours, 1, 15, 24, 26, 28, 60, 62, 84, 89, 95, 98, 118, 125

nest, 34, 51, 84, 99, 143

Nosema apis, 128, 137, 169

Nosema ceranae, 137

nucleus (nuc), 19, 46, 48, 60, 65–6, 68, 72, 82, 86, 90–1, 127, 146, 169

nurse bees, 51

oil seed rape (OSR), 80, 114

Omlet hive, 36–7

organic treatments, 83, 119, 141

organizations, 29, 43, 164

  urban beekeeping, 24

OSR *see* oil seed rape

packages of bees, 46, 48

permission, 28, 60

pests, 7, 123, 139–44

pets, 4, 28

pheromones, 9, 51, 58–9

piles of dead bees, 96

planning, 105, 110–11, 150

plants, 10, 53, 146–50, 158, 160

polishes, 1, 4, 127, 151, 156, 161

pollen, 3, 5, 10, 26, 51, 67, 77–8, 127–9, 159–60, 169

  basket 169

  pattie, 78, 123

  plants, 77, 146

  trap, 170

pollination, 1, 5–7, 32

population, 16, 80, 139, 147, 149

Practical Manual of Beekeeping, 157

Producing Royal Jelly, 161

products, 3–4, 7, 150, 156–7

propolis, 3, 11, 20, 50, 111, 118, 169, 170

pupa, 49, 140, 170

queen, 48–9, 50, 69–70, 73, 84, 121, 146, 147, 149, 150, 170

  cage, 170

  cells, 70, 74–5, 83, 88, 170

  clipping, 84–5, 170

  cups, 83

  excluder, 73, 74, 81, 88, 121, 123, 128, 147, 150, 170

  gentle, 100

  marked, 66, 71–2

  pheromones, 59

  re-queening, 100–1, 125

  substance, 59, 170

  virgin, 73, 75, 84

  young, 86, 125

Queen Bee, 159

Queen Rearing Simplified, 159

quilts, 41–2

ramp, 48, 62–3

re-queening, 100–1, 125

rearing, 17, 159, 169

recipes, 151–6

refractometer, 117

removable frame, 32–4, 38

reproduction, 83–4

robbing, 95, 97–9, 118, 148, 149, 170

rooftops, 24–5

royal jelly, 3, 51, 70, 73, 133, 170

running costs, 21–2

sahariensis, 54

sales, 22, 47

sampling, drone brood, 139

scientists, 9, 11

scouts, 51, 64, 92–3

secondhand, 22

self-spacing frames, 171

selling honey, 115

SHB (small hive beetles), 143

sheets, 34, 62–3, 166

sieves, 107

siting bees, 23–7
  rural, 26–7
  urban, 24–6

small colonies, 67, 74, 98–9, 148

smell, 53, 76, 132–3, 144, 156

smoker, 18, 19–20, 26, 45, 61, 78, 91, 171

snow, 124, 146

soiling, 28

specialist books, 127, 157–62

spiracles, 58, 142

spores, 135–7

spring, 5, 17, 52, 60, 62, 66–8, 70, 74, 76–8,
  80–1, 82–3, 127–30, 136–7

stamps, 1, 12, 160

statistics, 16–17

stings, 15–17, 44, 59, 171

Storch, H., 40, 95, 158

sugar syrup, 83–4, 122–3, 137

summer, 31, 49, 52, 78, 80, 82, 84, 86, 88,
  90, 92, 94, 96–8, 102–4

sun, 26, 55–6, 61, 118

supermarket honey, 3

supers, 39, 81, 85, 88, 110, 171

supplier, beekeeping equipment, 107

support, 17–18, 29, 36, 145

swarms, 19, 48, 125, 171
  artificial, 88
  catching, 90–3
  hanging, 92

starting with, 47–8, 55, 62–5

swarming, 83–93, 129–30, 148, 171
  preventing, 86–7
  signs of, 85

symptoms, 16, 132–4, 137, 142

tasks, 20, 51, 61, 67–8, 88, 104, 119, 120,
  123
  post-harvest, 119, 120
  pre-winter, 121
  wintering, 123

taxonomy, 53

temperature, 37, 53, 123–4, 143, 146

tfaya, 151

top bar hives, 18, 23, 38–41, 42–3

tracheal mite, 142

trees, 24, 26–7, 50, 91, 158

Tropilaelaps clarae, 141

UK's honeybee population, 8

ultra violet, 57

uncapping, 107, 112

uniting colonies/hives, 88, 89, 102, 149

urban beekeeping, 24–6, 84, 102

Varroa destructor, 67–8, 82–3, 138, 139
  control, 37, 83, 119, 129, 141
  monitoring, 139–41

venom allergy, 16–17

virgin queen, 73, 75, 84, 171

Virtual Beekeeping Gallery, 164

viruses, 42, 96, 138, 142, 169

waggle dance, 56

Warre hives, 39–40, 43

wasps, 1, 9, 17, 96, 117, 123, 143, 149–50

water, 25, 50, 76, 97, 117–18, 122
  content, 117
  source, 25–6

wax, 64, 108–9, 118–19, 125–6, 155–7
  comb, 64, 118–19, 133
  frames, 91
wax moth, 118, 134–5, 138, 172
WBC hives, 18, 32, 35, 47, 61, 98, 103
weather, 55, 66, 103, 128, 146–7, 149–50
websites, 35–6, 43, 45, 163–4
Western bee, 54, 67, 82
wings, queen's, 84
winter, 120–7, 131, 146–7, 149–50
  courses, 17, 45

projects, 125–6, 146
tasks, 123–4
thinking, 125
wire, 34, 100, 111
woodpeckers, 69, 144
worker bees, 49, 50–2, 58–9, 123, 172
  laying, 101–2, 168

year, 105, 125
young queens, 86, 125